NON-DESTRUCTIVE TESTING · SECOND EDITION

R. HALMSHAW

MBE, PhD, BSc, ARCS, C Phys, F Inst P,
Hon F Inst NDT, I Eng
Consultant

Edward Arnold
A member of the Hodder Headline Group
LONDON   SYDNEY AUCKLAND

Edward Arnold is a division of Hodder Headline PLC
338 Euston Road, London NW1 3BH

First published in the United Kingdom 1987
2nd edition 1991
Reprinted 1995

British Library Cataloguing in Publication Data
Halmshaw, R. (Ronald) *1919–*
   Non-destructive testing.—2nd ed.
   1. Non-destructive testing
   I. Title
   620.1127

ISBN 0 340 54521 6

Typeset in 10/11 pt Bembo by Wearside Tradespools, Fulwell, Sunderland
Printed and bound in the United Kingdom by St Edmundsbury Press,
Bury St Edmunds and bound by J W Arrowsmith, Bristol.

Preface

The industrial use of non–destructive testing (NDT) greatly expanded during World War II, although some of the techniques were known and used on a limited scale long before that date. It has attracted specialists, even specialists in one particular NDT method with only a limited knowledge in other fields of NDT, and this book has been planned to cover concisely the major fields of NDT and to indicate how they overlap.

It is written for engineers at undergraduate and graduate levels, to give an understanding of the physical principles, capabilities and limitations of a wide range of NDT methods, including the latest developments, so that the most appropriate techniques can be chosen for a specific application.

There is insufficient space in a book of this size to cover any NDT method exhaustively and an attempt has been made to provide a balance between the relative importance of the various methods.

Non–destructive testing is one part of quality control and therefore forms a link between the designer, the production engineer and the quality control department. Personnel in all three groups need to understand NDT and it is hoped that this book will meet this requirement.

I wish to thank friends and colleagues engaged in NDT in several countries for the opportunities I have been given for technical discussion on points of detail in the text, in particular Mr C A Hunt, my former colleague at RARDE.

I also thank my wife for her continuous support, her helpful criticisms and for typing the text.

R Halmshaw

Contents

1

Introduction

Non-destructive testing (NDT) has no clearly defined boundaries. A simple technique such as visual inspection is a form of non-destructive testing, as also might be the measurement of an obscure physical property such as Barkhausen noise. It used to be considered that there were five major methods— radiographic, ultrasonic, magnetic, electrical, and penetrant— but all these can be subdivided, and to them must be added a range of important new techniques such as acoustic emission methods, thermography, and holography. Further, a large number of even more specialised methods have been investigated and found to have applications in particular limited fields: these are methods such as proton annihilation, neutron scattering, proton radiography, microwaves, and nuclear magnetic resonance. In a book of this size it is not possible to describe in detail all the range of possible NDT methods.

Many NDT methods have reached the stage of development where they can be used by a semi-skilled operator following detailed procedural instructions, with safeguards built into the equipment. The advent of microcomputers allows procedures to be pre-programmed and cross-checked, so that a competent operator does not necessarily need to understand all the physics of the technique being used. However, it is desirable that the supervisors of the inspection, the designers who specify the techniques to be used in terms of their performance and attainable sensitivity, and the development engineers working on new methods, do have a thorough scientific understanding of the fundamental physics involved.

This book therefore is aimed at the undergraduate and graduate engineer, designer, or metallurgist, who needs to understand the basis of a wide range of NDT methods, so that the most appropriate one for a specific application can be selected. One of the problems in NDT is that there is often too large a choice of methods and techniques, with too little information, except sales literature, on the performance of each in terms of defect sensitivity, capital cost, speed of operation, running costs, or overall reliability. Some NDT methods have been much over-sold in recent years.

In this book, the amount of mathematics included has deliberately been kept to a minimum, partly so as not to burden the reader unnecessarily, and partly because although the mathematical principles of such topics as wave propagation or radiation scattering might be satisfying to an academic reader, they do not always add to the practice of NDT, nor do they necessarily point the way to future development.

There are, however, rapid advances being made in the computer-modelling of electrical, magnetic, and radiation fields, which appear to have very considerable potential in realistically representing the conditions met in practical specimens: where there has been useful progress, these methods have been mentioned in the appropriate chapters. At the end of each chapter, a list of references and further reading is included.

The terms 'non-destructive testing, NDT,' and 'non-destructive inspection, NDI,' are taken to be interchangeable, but a newer term 'non-destructive evaluation, NDE,' is coming into use. In NDT or NDI, in flaw-detection applications, the end product is taken to be a description of the flaws, which have been detected—their nature, size, and location. From this, either in conjunction with a standard for acceptable/rejectable flaws, or a knowledge of, for example, fracture mechanics, a decision is then made on the serviceability of the tested item. This decision is made by the designer, but in practice may be left to the NDT personnel, or the NDT inspector. In NDE, it is assumed that this acceptance/rejection of flaws is part of the non-destructive testing process.

There are only a few NDT applications, such as the testing of nodular cast iron, in which a direct relationship can be found between the NDT measurement and the strength of the material. With the development of data analysis by computer, however, together with pattern-recognition methods, it may soon be possible to analyse NDT data directly, in terms of component acceptability, so that the equipment can be programmed to produce (Go/No-go) decisions. These methods should therefore be described as NDE rather than NDT.

The terms 'flaw' and 'defect' have been used interchangeably and neither has been taken to signify either an acceptable or an unacceptable condition. More neutral terms such as 'discontinuity', 'imperfection', or 'inhomogeneity' are too cumbersome for general use, and the terms 'flaw sensitivity' and 'defect detectability' are so widespread in NDT, and have been used for so many years, that it seems unnecessary to propose anything different. It is understood, however, that for legal purposes the EEC has ruled that the term 'defect' signifies that the material, fabrication, etc., is defective, i.e. unserviceable. The term 'flaw' should therefore be used for any imperfections which are not considered to be rejectable. Thus, on this interpretation of the words, there is no such thing as an 'acceptable defect'.

The general topic of the determination of unacceptable defect sizes has been taken to be outside the scope of this book, as has the more general topic of quality control, of which NDT is obviously one part, but a short chapter is included on acceptance codes.

Although a great deal of non-destructive testing is carried out for flaw detection in materials—e.g. the detection of weld defects, lack of bond in adhesive joints, and fatigue cracks developing during service—it should not be forgotten that NDT has important applications in the examination of assemblies, to detect mis-assembled components, missing or displaced parts, to measure spacings, etc. In many of these applications, it is possible to be quite specific on what it is necessary to detect (i.e. the required sensitivity), or what accuracy of measurement is needed, and devise an NDT technique suitable for the particular application which may often be much faster or cheaper than a more conventional technique. Examples of this type are ordnance inspection (fuzes, etc.) for correction of assembly, aero-jet-engine inspection during test running to measure blade spacings during speed changes, ultrasonic thickness gauging, and metal alloy sorting.

There is a new factor coming into NDT, which seems likely to bring major modifications to most NDT methods. This is the use of computer techniques, using small computers. Apart from the obvious, rather trivial uses to simplify calculations, it is now possible to collect, store and process vast quantities of digital data at very high speeds. For example, in ultrasonic testing, in the signals produced by a transducer from a flaw there is a mass of data which is not used in conventional ultrasonic flaw detection. This can all be taken into a data store, and computer programs devised to extract information such as spectral composition, rise-time, pulse length, and maximum amplitude. At the moment it is not even certain, in some applications, what are the relevant properties of the signals which are needed. In addition, the computer can be used to choose the technique parameters for a given application, to adjust the equipment accordingly, and to provide warning if there are deviations, or a change in monitoring signals.

On NDT methods which provide an image, it is very likely that computers will be applied with pattern-recognition programs to interpret automatically the results of NDT. Perhaps a word of caution might be interjected here: the human eye is a very powerful and versatile instrument, when used with a trained memory, particularly in terms of pattern recognition against a noisy background, and the physics of imaging suggests that it might not be too easy to replace the eye with a computer. Against this, of course (computer) digital image enhancement techniques are already being used in a number of applications, particularly for signal-to-noise enhancement.

Returning to materials inspection, it has become fashionable in recent years to divide defect evaluation methods into quality-control criteria and fitness-for-purpose criteria. For the latter, acceptance standards must be defined on a case-by-case basis, usually using fracture mechanics. For the quality-control criteria, the requirements are based on a more general engineering experience, and the inspection is directed towards detecting the most common manufacturing defects, with an implication of less-severe NDT requirements. These concepts again emphasise the need, not always fully documented, of having a knowledge of the defect sensitivities required, in detail, before applying NDT.

Regarding the comparison and evaluation of NDT methods, there have been several published papers purporting to compare the performance of different NDT techniques and methods. Nearly all these reports have taken a particular type of defect or specimen, collected or fabricated a set of specimens with different sizes or severities of the defect, inspected them non-destructively, and then analysed the results statistically. Few of these trials have contained enough specimens to be statistically significant when there may be problems of reliability and measurement accuracy as well as defect detectability. More importantly, the results obtained, although valuable, depend almost entirely on the nature of the particular defect used. Thus, in a trial involving small fatigue cracks at section-changes on a light-alloy panel, it is not surprising that eddy current and penetrant testing were superior to ultrasonic testing and radiography. On corroded surfaces, penetrant inspection was found to be inferior to the eddy current technique, in reliability if not in sensitivity. A comparison of ultrasonic and radiographic methods will produce very different conclusions if the defect chosen is porosity rather than cracks.

The use of a minimum detectable defect size in any specific application is not an efficient assessment of either a technique or an operator. Probabilities and confidence levels are also needed, and laboratory trials cannot necessarily be extrapolated to field results. Most NDT techniques have a wide range of applications and comparison data is valid only for a particular application, a specific type of defect, and a particular material.

SI units have been used throughout the text, except where US Standards are quoted. Many of these still use inches, and in these cases the original American wording has been retained. If a standard asks for a $\frac{1}{16}''$ hole, it seems unnecessarily pedantic to convert this to 1.588 mm. Micrometres (μm) have been used in preference to microns, and 'mils' has not been used.

This book has been arranged with each major NDT method presented separately, and in this second edition short chapters on 'Computers in NDT' and on 'Reliability and Probability' have been added. The sections on eddy current testing and acoustic emission methods have been considerably expanded.

A case could also be made that some applications of NDT should be treated as separate subjects: thus the measurement of stress in metals now includes a wide range of NDT methods, ranging from the classic X-ray diffraction technique, through a range of ultrasonic methods such as velocity change, polarisation effects, to magnetic methods such as Barkhausen noise, magneto-elastic and magnetoinductive methods, to thermoelastic methods. A useful review and bibliography on stress measurement is given by Hauk in reference 1.

Since the first edition of this book the use of computers in NDT has increased, and another development is the proliferation of acronyms. Much use is made of acronyms in computer software, and it seems that almost every new instrument, particularly in the ultrasonic field, must have an acronym as a name. As far as possible these have been explained, and the number of them used reduced to a minimum.

Further reading: reference

1. Hauk V, *Proc 4th Europ Conf NDT (London)*, **3**, 1580, Pergamon Press, Oxford, 1988

2

Visual methods

2.1 INTRODUCTION

Visual methods of surface examination, with or without optical aids, tend to be neglected by NDT personnel but are, nevertheless, important. Many of the most serious defects, from the strength point of view, are surface-breaking, and while the detection of these can be enhanced by magnetic particle inspection, they can also often be seen by careful direct visual inspection. Such weld defects as severe undercutting, or an incompletely-filled groove, can be easily seen if the surface is accessible, and can lead to immediate rejection or rectification without the need for more expensive testing by ultrasonic methods or by radiography. In addition, surface shapes (contour gauging, profile gauging) and surface roughness provide valuable quality-control information. On a microscopic scale, the preparation of a local area of a specimen together with metallographic examination at various magnifications is a combination of destructive and non-destructive inspection, but often the amount of metal removed in preparing a local area of surface for examination is not sufficient to affect the subsequent serviceability of the component. In addition, surface replicas can be obtained both for macroscopic and microscopic examination. Finally, high-speed surface inspection, with automated output, is in itself a non-destructive inspection method for such products as bright steel sheet (see Section 2.5).

2.2 OPTICAL AIDS

Aids to visual inspection should be used whenever practicable. For local examination of a portion of a metal which is directly accessible, a small hand lens, used in conjunction with a portable light such as a pen-torch, is very useful. A magnification of ×2–×4 is all that is usually required.

Industrial telescopes, usually known are borescopes, or introscopes, enable surfaces inaccessible to the naked eye to be seen. They are best known for examining the internal surfaces of tubes and piping, and for inserting into access holes in machinery. An important application is to look at the surface of turbine blades in an aero-engine for cracking and corrosion, without dismantling the engine.

The design of borescopes has been transformed in recent years by the application of optical–fibre techniques. These can illuminate a surface and retrieve the image over distances up to several metres. The desirable properties of a borescope are that it has as large a field of vision as possible, minimum image distortion, and adequate illumination. Borescopes are now available which are coupled to a closed–circuit television camera, to provide an image on a monitor screen. As an alternative technique, sub–miniature television cameras have been built for direct insertion into small pipework. If a television camera is used, the image can be taped, or disc–recorded. A further variant is to insert a small film camera into the structure and by remote positioning and operation, photograph the relevant areas directly on film. This has been carried out in radioactive environments in a nuclear power plant, by using film which is highly sensitive to light, but which is not over–exposed by the background ionising radiation during positioning and exposure. A pro-grammable manipulator with a miniature television camera is used to set up the positions and lens functions for a series of locations and then the TV camera is replaced with a film camera for a very rapid sequence of exposures.

In all these remote viewing systems, the two main problems are:

(1)   the effect of the direction of the illuminating light—if possible, this should be capable of being varied so that detail can be seen in relief, and glare and dazzle effects eliminated,
(2)   identification of the precise areas being seen.

2.3   IN-SITU METALLOGRAPHY

This is not widely used, perhaps because of a lack of awareness of its potential value. A local area can be ground, polished, and etched. Electrochemical etching is easily done, in situ, even on a vertical surface, using a glass tube and a sealant to the surface. A portable microscope with a camera attachment is then used on the prepared area. Alternatively, a replication method can be used with cellulose acetate films moistened with acetone and spread on the etched surface. When dry, these films are peeled off and held between glass strips: they are examined with a reflecting aluminium foil background.

An alternative method of producing a replica is to use a varnish having a nitro–cellulose or plastic base, spread on with a spatula, and allowed to dry. Great care is needed in lifting the replicas off the surface. Instead of the aluminium foil background for viewing, the surface of the replica not containing the impression can be made reflecting by vacuum deposition of an aluminium coating.

Often a study of the surface microstructure of a material, in situ, can provide additional information to supplement other NDT findings.

2.4   OPTICAL HOLOGRAPHIC METHODS

The advent of lasers and the science of coherent optics has led to a whole new range of optical inspection techniques, which come broadly under the heading of visual inspection. In addition, the ability to collect an optical signal with a television camera, digitise and store the data, and then subject the data to digital image processing, has further extended these techniques.

A group of techniques under the general heading of optical holography can be used for the comparison of specimens, or for the measurement of small amounts of deformation under stress, or for a study of the surface during vibration. As the presence of a defect is likely to cause changes in the deformation or vibration pattern, the methods can, by extension, be used for defect detection.

2.4.1   Principles of holography

If a specimen is illuminated with light, different parts of the surface scatter the incident light, producing light waves which have an amplitude and relative phase. Normal image storage systems such as photographic film respond only to light intensity (amplitude)^2 and do not utilise the phase information.

In conventional holography the light scattered from the surface of a specimen illuminated with coherent (laser) light is made to interfere with another reference field of light derived from the same laser. This transforms the phase information into intensity information, so that all details of the distribution of amplitude and phase can be recorded on a photographic film (the hologram) if this has adequate resolution. A pattern of light and dark fringes is produced. After the photographic plate P (Fig. 2.1(a)) is chemically processed it can then be used to diffract the reference beam of light (Fig. 2.1(b)) to produce a three-dimensional image.

The pattern on the hologram acts like a complex diffraction grating, and an observer looking through the hologram plate sees the original object in place, even though it has been removed.

The image is a virtual one but is three-dimensional, and if the observer moves his head sideways there is a full effect of perspective and depth. If this image is reconstructed with the object in its original position, the three-dimensional image superimposes exactly on the object, but if the object has moved slightly or been deformed locally, the observer sees bands of interference fringes on the surface, the number and spacing depending on the amount of object movement. This is holographic interferometry.

The following fundamental points about optical holography can be made.

(1)   The light source must have suitable coherency properties, which in practice means laser light. A typical He–Ne gas laser has a useful temporal coherence length of about 20 cm.

(2)   Very stable conditions are necessary during exposure-time. The relative motion between the object and photographic plate should be less than $\lambda/4$, where λ is the wavelength.

(3)   If the angle between the reference beam ray and any one scattered ray from the object is β, then the fringe spacing, δ, is given by

$$\delta = \lambda/\sin\beta$$

so for $\beta = 30°$, $\delta = 2\lambda \simeq 1\ \mu\text{m}$, assuming light of $\lambda \simeq 0.5\ \mu\text{m}$.

Thus the photographic emulsion must have a very high resolution if there is to be a useful field of view. A resolution of 1000 lines per mm is desirable, and

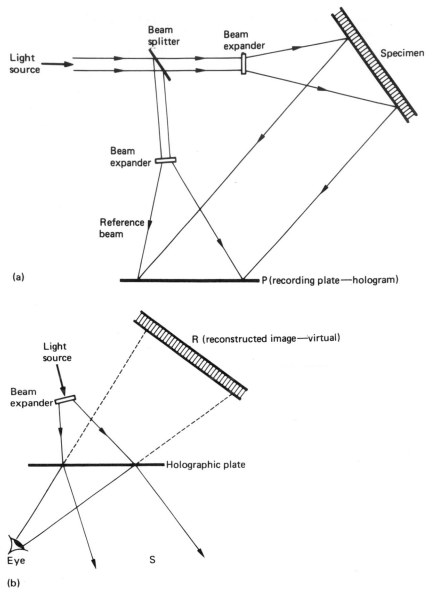

Fig. 2.1   Optical holography: (a) hologram formation and (b) image reconstruction

this in turn requires a very high light intensity to maintain reasonable exposure-times.

By taking a hologram of a stationary surface, replacing the plate in its original position so that the holographic image is superimposed on the object, and then observing the moving fringes produced when the object is vibrating, there is effectively a 'live fringe' effect. The motion can be arrested by using pulsed illuminating beams at the reconstruction stage and pulsing these at the same frequency as the vibration. If the pulse-time is kept short, high contrast fringes can be seen, even at high orders of interference.

While three-dimensional holographic images can be very impressive and also useful, e.g. in displays of museum specimens, the main use in NDT is to detect minute deformations by holographic interferometry. One technique is to expose the holographic plate for half the required exposure-time, and then apply a small strain to the specimen mechanically, or by heating, for the other half of the exposure.

For internal defect detection in solid specimens by deformation, a useful rule-of-thumb for holographic testing is that the defect should be at least twice as large in diameter as it is deep, or that a crack should be longer than the material thickness.

In this method of double-exposure holographic interferometry, let the amplitude of the initial wave at position (x, y) on the holographic plate be $U_1(x, y)$, and that of the second wave after a small object movement be $U_2(x, y)$, where

$$U_1(x, y) = a(x, y) \exp[-i\phi(x, y)]$$

and $$U_2(x, y) = a(x, y) \exp\{-i[\phi(x, y) + \Delta\phi(x, y)]\}$$

a being a constant and $\phi(x, y)$ the phase component at (x, y). Then dropping the (x, y) for simplicity of notation, the resulting intensity, I, from addition of the waves is

$$I = |U_1 + U_2|^2$$
$$= |a \exp[-i\phi] + a \exp\{-i[\phi + \Delta\phi]\}|^2$$
$$= 2a^2\{1 + \cos[\Delta\phi]\}$$

This then represents the original brightness, crossed by fringes of spacing

$$2\{1 + \cos[\Delta\phi(x, y)]\}$$

That is, any local deformations or distortions will be revealed as local fringes. Holographic interferometry is more suitable for detecting the bulging or swelling of a surface, rather than for surface stretching.

A second technique is time-averaged holographic interferometry, which is usually applied to a vibrating surface. If a hologram is made of the vibrating surface, the reconstructed image will be found to contain a pattern of interference fringes, and by taking holograms at different frequencies, the various modes of vibration can be determined.

Real-time holographic interferometry is similar in principle to the double-exposure technique, except that the object forms one of the two waves—for example, $U_2(x, y)$. The fringe patterns produced follow the motion of the object. It is necessary to expose, process, and replace the holographic plate, and its repositioning must be accurate to within one-quarter of a micrometre. Thermoplastic holographic plates which can be processed in situ are now available. The use of pulsed solid-state lasers, with much higher brightnesses, can obviate the need for the extremely stable mountings needed for continuous lasers, by reducing the photographic exposure-time to a fraction of a second. However, such lasers have a shorter coherence length.

2.4.2   Applications of optical holography

Most of the NDT applications have been in the aerospace field, e.g. for composite panels, honeycomb structures, and bonds and disbonds in adhesively-bonded structures. The specimen is given a slight stress either mechanically or thermally, and non-uniform effects due to defects are shown by fringe anomalies. Holographic NDT is particularly useful on complex shapes where other NDT methods, such as ultrasonic testing, are difficult or time-consuming to apply. Successful applications in the vehicle tyre industry and to printed-circuit-boards have been reported.

Using a television camera and digital recording, the amplitude and phase information can be obtained in near-real-time, a technique sometimes called 'electro-optical holography' (Stetson & Brohinsky, 1985).

2.4.3   Speckle

The properties of coherent radiation may be used directly for non-destructive testing without recording a hologram. If laser light is scattered from an object, the surface appears speckled: that is, it appears to be covered in fine light and dark areas which move as the eye is moved. A qualitative explanation is that each element of speckle represents the smallest area which the eye (or an optical system) can just resolve, and since this area may be quite large and irregular, compared to the wavelength, the light scattered from it will be made up of a number of waves with random phase differences. These waves interfere with one another to produce a resultant intensity which can vary from zero to a maximum value. Statistically, the distribution of the intensities of resolvable areas is random, within these limits, and so each area is likely to have a different brightness to its neighbour, so producing the speckle effect.

The speckle pattern is therefore related to the detailed surface structure (and to the resolving power of the system used to view or record it). Reducing the resolving power increases the apparent size of the speckle.

Speckle can therefore be used to detect movement of a surface and to detect fatigue effects causing surface distortion, by recording the speckle image of a test surface on a photographic plate. After processing, the plate is replaced in exactly the same position and acts as a negative mask, so that correct replacement gives a uniform minimum transmitted field. If the surface now moves, due to the development of fatigue defects or other causes, the matching is no longer perfect, transmission is increased, and a signal is recorded.

2.4.4   Electronic speckle pattern interferometry (ESPI)

Speckle patterns can be used as an inspection technique, without the need for photography, by using a closed-circuit television (CCTV) camera.

The television camera system should preferably have a digital output and be equipped with a digital framestore, a choice of filters (bandpass video), and the means for image subtraction. The set-up of the equipment for vibration mode viewing, or surface displacement measurement, is shown in principle in Fig. 2.2. The procedure is to adjust the system aperture and the filter to produce the best speckle uniformity on a stationary object. The object is then

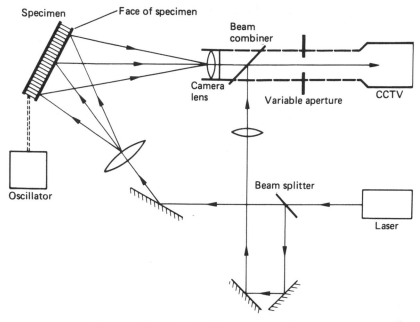

Fig. 2.2 Electronic speckle pattern interferometry

vibrated at resonance and the electronic filter is tuned for optimum fringe visibility, when the spacing of the fringes can be analysed exactly as with time-averaged holography. The nodal areas and vibration amplitudes can be determined.

The dynamic range of ESPI is not as good as with holographic plates, and usually only a few fringes can be seen, but this can be improved if the reference beam is modulated by inserting a vibrating mirror in the beam with the same frequency of vibration as the specimen. In addition, digital image processing can be applied to optimise the contrast of the fringe images and to detect the zero-order fringe position. Use of a reference-beam phase-stepping technique from a digitisation of the interference patterns, allows the calculation of the phase changes without the need to produce fringe patterns.

A further development of ESPI is to use a two-wavelength laser to compare specimen shapes, independently of variations in surface finish.

Much of the future development in both holographic and speckle techniques depends firstly on the availability of higher-powered lasers, with long coherence lengths, and secondly on the availability of photographic plates of very high resolution and adequate sensitivity to the laser 'light' being used. A carbon dioxide laser is attractive in that it emits a lot of radiation, but the wavelength is 10.6 μm (infrared) where photographic materials have very low sensitivity. Pulsed lasers overcome the stability problems by their short pulse length (typically 30 ns), but tend to have short coherence lengths which prevent them being used on large specimens.

Instead of light, neutrons can be used for holography and this technique is briefly described in Chapter 9. Acoustic holography is described in Chapter 4.

2.5   DYNAMIC SURFACE-INSPECTION

Products with surface defects such a holes, dirt, scratches, can cause serious manufacturing problems, particularly when a subsequent surface treatment is needed. Manual inspection has serious drawbacks such as subjectivity, low speed and high cost.

The use of television cameras to enhance direct visual methods has already been mentioned, but a quite distinct group of applications uses the electronic signal from a television camera to produce an automated system operating at very high inspection speeds. Usually, the television signal is converted into digital form, when it can be stored and used in conjunction with a computer program to search for particular anomalies in the signal which originate from surface 'defects'.

A very wide variety of applications has been described, such as the detection of scratches on a hydraulic cylinder surface, porosity on car brake discs, chips and cracks in the necks of glass bottles, and warped rims on crockery.

In some cases, light detectors with fibre-optic coupling are used instead of television cameras and inspection speeds of 60 000 items per hour have been quoted. Image sensing devices such as photodiodes and charge coupled devices, CCDs, are also used as self-scanning arrays, with a digital shift register for very-high-speed optical scanning. CCDs are available as 800×800 element arrays with a 25 μm spacing, and together with a lens can be used in place of a television camera, to provide a digital output (solid–state television camera). Such line sensors usually provide a higher resolution than a television camera. For direct light detection, the usual method is to use a scanned linear array.

Photodiodes, CCD elements, and charge injection devices, CIDs, all depend on the optical sensitivity of silicon. In silicon, photons within a certain energy range can knock electrons out of atoms, leaving positive 'holes'. The holes and electrons have some mobility before being destroyed by recombination. The electrons can be collected by an electric field at the diode junction between two different forms of silicon (n-type and p-type). A clock connects each sensor in turn with a video-line (Fig. 2.3), and the scanning function is formed by an analogue shift register which moves the charge packets created by light through the silicon chip. Most CCDs have photodiodes as sensors feeding a CCD register and are usually described as 'charge coupled photodiode arrays', but for the highest light sensitivity and very-high-speed operation, CCDs without the photodiode layer are necessary. CID (charge injection devices) are also used for two-dimensional detectors. These consist of a combination of digital scanning and MOS-type sensing elements. All silicon-based arrays have some direct X-ray sensitivity, but are easily and progressively damaged by ionising radiation.

As an example of an application of a light scanning inspection system, measurement of specular reflection from the surface of a steel sheet has been used for the quality control of surface finish.

In one arrangement (Fig. 2.4), a two-dimensional scan is obtained by moving the steel sheet at a constant velocity and also using a rotating polygon mirror to scan across the surface. A cylindrical lens is used to adjust the optical path length, so as to obtain a uniform value following reflection of the light beam from the mirror. A laser source is used for illumination and a stationary

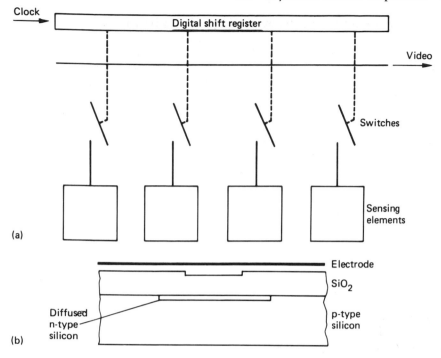

Fig. 2.3   (a) Self-scanned photodiode (SSPD) array. (b) Enlarged section of sensing element

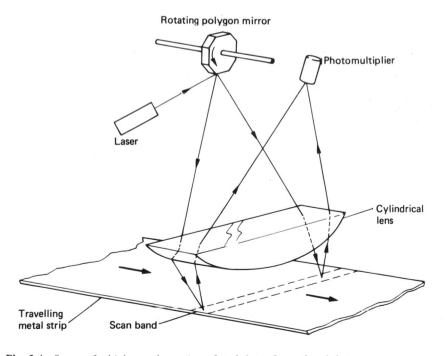

Fig. 2.4   System for high-speed scanning of steel sheets for surface defects

photomultiplier detects the scattered light. If there is a defect on the surface under the light beam, the scattered intensity increases; if the defect is a crack, it decreases, from the background level. The detector output can be displayed on a video screen or recorded as a printed number. In one system, the scanner acquires 10^4 data points per ms on a 3 mm^2 area.

Another method combines specular and non-specular measurements using two directions of surface illumination to eliminate dependence on surface roughness.

There are two main types of local defect. If a defect is seen as an area of surface darker or lighter than the area around it, it can be called a 'contrast-defect'. On textured or patterned surfaces there can be defects which are close to the background brightness, but which cause a change in pattern or texture: these are 'pattern defects'.

Whether the signals are collected by a line sensor or an area sensor, digital image processing is generally used to analyse the data. This may range from simple thresholding, thresholding after pre-filtering, two-dimensional convolution for edge-detection, or the use of various kernels such as Sobel or Laplacian operators.

The detection of pattern defects on a textured surface depends on prior knowledge of the pattern. The use of a high-pass filter to remove gradual intensity changes, followed by thresholding, is usually successful.

Once a defect is detected it has to be assessed for severity so that a pass/fail decision can be made; usually both area and contrast contribute to severity. The sum of the contrast of pixels within the defects can be used as a measure of severity, and this can be built into a computer inspection program.

There are many applications of optical techniques for surface inspection, such as shadow moiré, photogrammetry, stereoscopic viewing, and holography, which are on the fringes of non-destructive testing. Their industrial application depends largely on the nature of the surface, many real engineering surfaces being non-specular and also curved.

Further reading: references

Batchelor B G, Hill D A and Hodgson D C (eds), *Automated Visual Inspection*, IFS Publications Ltd, 1985

Butters J N, *Engineering Uses of Coherent Optics*, Cambridge University Press, 1975

Erf R K (ed), *Holographic Non-destructive Testing*, Academic Press, New York, 1973

Erf R K (ed), *Speckle Metrology*, Academic Press, New York, 1978

Sollid J E, Optical holographic interferometry, Chapter 7 of *Research Techniques in Non-destructive Testing*, Volume 2, Sharpe R S (ed), Academic Press, 1973

Stetson K A and Brohinsky W R, Electro-optical holography, *Applied Optics*, **24**, 3631, 1985

Williams D C and Robinson D W, Electro-optical holography, *Proc 4th Europ Conf NDT (London)*, **1**, 588, Pergamon Press, Oxford, 1988

3

Radiological methods

3.1 SCOPE

Ever since the discovery of X-rays in 1895, it has been realised that they can be used for the non-destructive testing of materials as well as people. There are records of 'industrial' radiographs being taken as early as 1896.

X-rays are a form of electromagnetic radiation, of the same physical nature as visible light, radiowaves, etc., but which have a wavelength which allows them to penetrate all materials with partial absorption during transmission. They include a fairly wide waveband of radiation from about 10 nm, usually called Grenz-rays or 'soft' X-rays, which will usefully penetrate only very small thicknesses of solid material, to about 10^{-4} nm, often called high-energy or 'hard' X-rays, which will penetrate up to about 500 mm steel. Gamma-rays, emitted by radioactive sources, can exist over a similar wavelength range; they too are electromagnetic radiation with exactly the same properties and are also widely used in industrial radiography. Neutrons of suitable energy have similar properties of partial absorption in materials and so can be used for radiography: although neutrons are atomic particles and not electromagnetic radiation, 'neutron radiography' is a well-established term and technique. Protons have also been used for radiography.

X-rays and gamma-rays travel in straight lines outwards from a source: for all practical purposes they cannot be focussed, so the usual set-up for producing a radiograph is as shown in Fig. 3.1, using a small diameter source, G, and a sheet of photographic film as detector. By conventional, perhaps rather pedantic, definitions, a *radiograph* is 'a photographic image produced by a beam of penetrating ionising radiation after passing through a specimen', and *radiography* is 'the production of radiographs'. Although today the recording sheet in the figure need not be a photographic material, most people would probably still refer to the result as a radiograph.

Instead of using a photographic layer at plane AA' (Fig. 3.1), the X-rays can be absorbed in a layer of fluorescent material which converts the X-rays to light: this technique is called 'fluoroscopy' or 'radioscopy', but newer terms, such as 'real-time radiography, RTR', are widely used. If the X-ray image at plane AA' is detected and measured with some form of radiation detector, such as an ionisation chamber, Geiger–Müller counter, or scintillation counter, the technique is usually known as 'radiometry'.

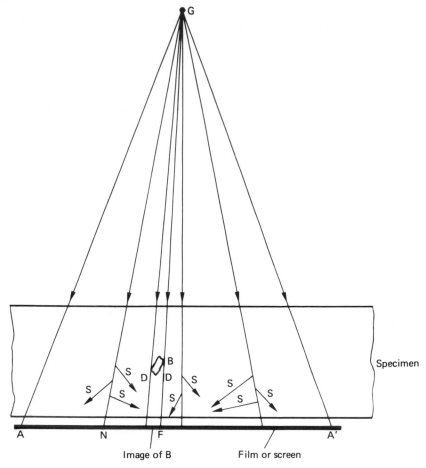

Fig. 3.1 Principles of film radiography, where AA′ is the film plane, B is a cavity in the specimen, imaged at F, D is the direct, image-forming radiation, G is the radiation source, and S is the scattered radiation generated inside the specimen

X-rays can also be used to produce diffraction patterns to study material structures. These techniques of diffraction and X-ray fluorescence are usually regarded as outside the scope of industrial radiology.

X-rays and gamma-rays are hazardous, and suitable safety precautions must be used when they are employed: the general problems of radiation safety are outlined in Section 3.12.

3.2   GENERAL PRINCIPLES

Returning to Fig. 3.1, the X-rays travel in straight lines from the source to the film, so that if there is a cavity in the specimen, as shown at B, which causes a lower absorption along the path GBF, more radiation reaches the film at point F, compared with (say) point N, and an X-ray 'image' of the cavity is produced which will be the projection of the cavity, very nearly natural-size (very slightly enlarged). This is therefore a two-dimensional image of a three-dimensional cavity.

To produce a radiograph, the X-rays are allowed to reach the film for an appropriate exposure-time, which depends on the intensity of the X-rays, the thickness of the specimen, and the characteristics of the film. The film is then chemically processed (developed, fixed, washed and dried), so that the X-ray image can be seen as different levels of blackening (film density); the film is then placed on an illuminated screen so that the image can be examined and interpreted.

When X-rays are absorbed in material, some X-ray energy is re-emitted as scattered radiation, which under some conditions can travel in a different direction to the primary beam. Thus at point F on the film, some radiation will travel directly along the line GBF, and this forms the image of the cavity B, but scattered radiation, S, can also reach F, and this is non-image-forming. The ratio of intensities

$$\frac{\text{image-forming radiation}}{\text{non-image-forming radiation}} = \frac{\text{direct radiation}}{\text{scattered radiation}} = \frac{I_D}{I_S} \tag{3.1}$$

where I_D and I_S are the intensities of the radiations, is an important parameter in industrial radiography.

The total radiation reaching point F in unit time is

$$(I_D + I_S)$$

and the ratio

$$\frac{I_D + I_S}{I_D} = 1 + \frac{I_S}{I_D} \tag{3.2}$$

is known as the 'build-up' factor.

The basic law of X-ray absorption is given by

$$I_x = I_0 \exp(-\mu x) \tag{3.3}$$

where x is the thickness of the material, I_0 is the incident intensity of radiation, I_x is the transmitted intensity, and μ is a constant known as the 'linear absorption coefficient' with dimension cm^{-1}, and its value depends on the material and the X-ray wavelength.

Equation 3.3 is strictly true only for monoenergetic radiation, and for narrow-beam conditions under which the amount of scattered radiation reaching the detector is small enough to be neglected, but it is often also applied to broad-beam conditions and mixed-energy radiation beams, to determine an 'effective' value of μ related to practical radiographic conditions.

Two useful derivatives of μ are widely used:

(1)   the half-value thickness, HVT (the thickness of material required to reduce the intensity by a factor of two),

$$HVT = 0.693/\mu \tag{3.4}$$

(2)   the tenth-value thickness $= 2.303/\mu$. $\tag{3.5}$

As most X-ray and gamma-ray beams are not monoenergetic, the energy content of the beam changes as it passes through the specimen thickness, so that the practical value of HVT and $\frac{1}{10}VT$ are not constant with thickness and are not the same as the values determined using the true (narrow–beam) value of μ.

One other important physical factor should be mentioned at this stage. The source of X–rays or gamma–rays is usually a small area, a few millimetres in diameter: it is never a true point source. Consequently, for any defect in the specimen which is not close to the film, there is a blurring of the image, known as the penumbra, or geometric unsharpness, U_g (Fig. 3.2). In this figure, the size of the source, s, has been exaggerated, for clarity, and the 'defect' has been taken to be a small, physically sharp step on the surface, at O. Then an element of the source at M will image the step on the film at M' and an element of the source at P will image at P'. Thus the image of a sharp step is

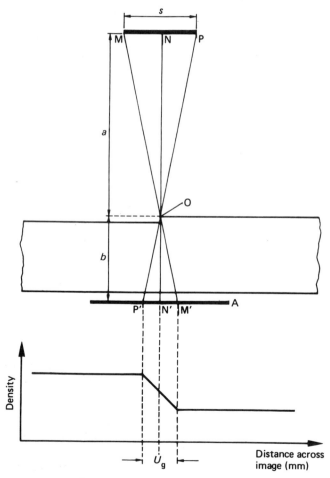

Fig. 3.2   Geometric unsharpness on specimen detail at O, with a curve of the density distribution across the image (source diameter exaggerated in size)

spread out—blurred—on the film, over a distance $d_{P'N'M'}$. By simple geometry

$$\frac{d_{P'M'}}{d_{MP}} = \frac{d_{N'O}}{d_{ON}}$$

and if for simplicity $d_{NO} = a$ and $d_{N'O} = b$,

$$d_{P'M'} = \frac{sb}{a} = \text{geometric unsharpness, } U_g \qquad (3.6)$$

and this relationship holds for the image of the edge of any specimen detail.

   This treatment of the focal spot of diameter s, as a uniformly emitting area, is an over-simplification of the real situation. Practical focal spots are non-uniform emitters so that the effective size of the geometric unsharpness and its influence on the flaw imaging is a more complex problem. The problem becomes increasingly important with microfocus X-ray tubes, where the emitting area on the target may be much wider than the 'true' focal spot size.

   There are other causes of image unsharpness in radiography, so for the 'geometric' cause the suffix 'g' is generally used. The implications of geometric unsharpness will be described in more detail in Section 3.7.

   A second aspect of geometric image formation is that while a cavity in a specimen is three-dimensional, its image on the film is two-dimensional. If a cavity is spherical, the image will be circular irrespective of the direction of the radiation (Fig. 3.3(a)), but, if the cavity is disc-shaped or planar, the image depends very much on beam direction (Fig. 3.3(b) and (c)). Thus a crack imaged 'edge-on' will be seen as a sharp, narrow line (Fig. 3.3(b)), whereas an oblique crack will be imaged as a broader, much fainter band (for example, width d in Fig. 3.3(c)), rather than as a narrow line.

3.3   IONISING RADIATION

X-rays are produced when high-energy electrons strike a metal target, and they originate in the structure of the atoms of the target. Gamma-rays are emitted from the nucleus of some radioactive elements. Both X- and gamma-rays travel at the speed of light, travel in straight lines, and are invisible. They have a photographic action similar to that of visible light.

   Planck's quantum theory postulates that electromagnetic radiation is not radiated or absorbed continuously, but can be regarded as small packets of energy, which he called *quanta* or *photons* and the energy, E, of each quantum is given by

$$E = h\nu = hc/\lambda \qquad (3.7)$$

where h is Planck's constant (6.63×10^{-34} J s), ν is the frequency of the radiation, and λ is the wavelength of the radiation.

   The wavelength of X-rays is usually quoted in nanometres (nm). The energy of an X-ray quantum, E, can be given in electron-volts (eV), where $h\nu = eV$, e is the charge on the electron, and V is in volts. Similarly, the energy

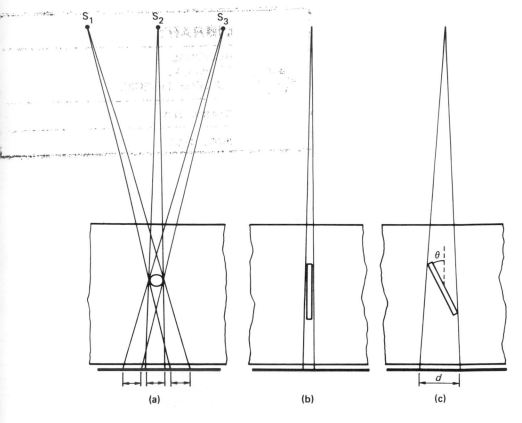

Fig. 3.3 Shapes of image for different defects: (a) spherical or cylindrical cavity (image is the same shape whether source is at S_1, S_2, or S_3), (b) planar slit or crack in plane of X-ray beam, and (c) planar slit or crack at angle θ to X-ray beam

of a gamma-ray quantum can be given in keV or MeV, even though no electron is involved in the production.

When an electron having some kinetic energy strikes the target of an X-ray tube there is a transformation of energy. The electron is rapidly decelerated in the electric field of the nucleus because of its small mass, and a quantum of X-radiation is emitted with a minimum wavelength given by

$$\lambda_{min} = \frac{hc}{eV} = \frac{1239.5}{V} \text{ nm}$$

Part of the electron energy may be used in removing an electron from a target atom, before the incident electron is stopped by direct interaction with the nucleus, so that a number of X-ray quanta with different wavelengths are produced, i.e. a continuous X-ray spectrum (Fig. 3.4). By analogy with visible light, such an X-ray spectrum is sometimes called 'white radiation' or *bremsstrahlung*. The precise shape of the curve depends on the target material and thickness, as well as on the energy distribution in the incident electron beam.

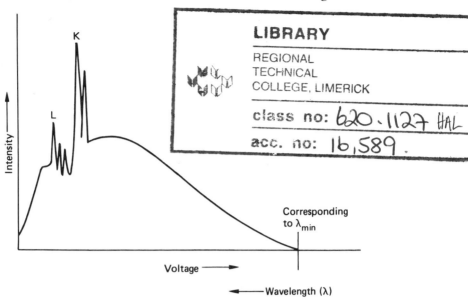

Fig. 3.4 Intensity distribution in an X-ray beam, plotted against voltage and wavelength

As the energy of the incident electrons is easily measured in terms of the voltage difference across which the electrons are accelerated, it is common practice in industrial radiology to describe an X-ray output such as that shown in Fig. 3.4 as, for example, 200 kV X-rays, where 200 kV corresponds to λ_{min}. Even with a high-energy X-ray machine in which there is no direct generation of a high voltage, the X-ray output is usually described in the same way, e.g. 20 MV X-rays. The use of '200 keV' would refer to a single-energy X-ray quantum and not to the output from an X-ray tube.

In Fig. 3.4 there are some low-energy, sharp-intensity peaks. These are the K, L *characteristic X-rays* whose energy depends on the material of the target. When an electron is removed from a target atom, the atom is left in an 'excited' unstable state and an electron from another electron orbit tends to jump into the vacant space; the difference in energy between the two electron orbits has a series of fixed values which produces the characteristic X-rays in the K, L bands. The shortest wavelength K-radiation (uranium, $K_{\beta2}$) has a wavelength of 0.011 nm (i.e. an energy of 0.11 MeV). This 'characteristic radiation' is of minor importance in most industrial radiography, but of considerable interest in X-ray diffraction and spectrometry work.

3.3.1 Scattered radiation

The absorption of X-rays is not always a simple process in which the absorbed X-ray energy changes to some other form of energy and disappears. There are several types of X-ray absorption process, but only four are of concern in industrial radiology and the total absorption is the sum of these. Each of these four is predominant in different parts of the X-ray energy range.

In *Rayleigh scattering*, the incident X-ray photon is scattered with no change in the internal energy of the scattering atom, and no release of electrons. There is a definite phase relationship between the incident radiation and the scattered radiation, so that the scattering is coherent. Broadly, Rayleigh scattering is important only below 0.1 MeV, and even in this region is only a minor contributor to the total attenuation.

In *photoelectric absorption*, the X-ray photon is absorbed and its energy is used in removing an electron or electrons from the outer shells of the absorbing atom. The consequent rearrangement of the electrons to fill the gaps produces the characteristic X-rays already referred to. If the incident electron has sufficient energy to remove an orbital electron completely from the atom, the atom is said to be *ionised* and the electron released is given some kinetic energy. Photoelectric absorption is a dominant effect in the low-energy region, up to about 0.5 MeV, particularly for high atomic number absorbers.

In *Compton scattering*, the incident X-ray photon is scattered non-elastically and a recoil electron is also produced out of the absorbing atom. With most of the X-ray energies used in industrial radiology, this is the dominant contributor to total attenuation. The result of this scattering process is a secondary photon of a lower energy than the incident photon, travelling in a different direction, and a recoil electron (Fig. 3.5). As the primary photon energy increases, so the recoil electron's direction, ϕ, becomes closer to zero.

For a single scattering process, the Compton relationships are:

$$\lambda = \lambda_0 + \frac{h}{mc}(1 - \cos\theta)$$

$$\cos\theta = 1 - 2\left[\left(1 + \frac{1}{\lambda_0}\right)^2 \tan^2\phi + 1\right]^{-1} \tag{3.8}$$

Thus, as the incident energy increases, the scatter tends to be more in the forward direction and also to be higher in energy, closer to the energy of the incident photon.

In practical radiography, where a photon passes through a large thickness of absorber, multiple Compton scattering occurs, which is exceedingly difficult to compute accurately. Some calculations have been made by using Monte Carlo methods. In most practical radiography, this scatter, being non-image-forming (see equation 3.2), reduces the image quality, but there are applications which deliberately make use of scattered radiation (see Section 3.11).

Pair production occurs only if the primary photon has an energy greater than 1.02 MeV, and it is not really important until much higher energies are reached. In this process an incident photon disappears, with the creation of an electron–positron pair. The positrons have a very short life and their energy reappears as two photons of energy 0.5 MeV travelling in opposite directions. The electron can be reabsorbed, producing bremsstrahlung, so that absorption by pair production results in scattered radiation of mainly 0.5 MeV energy. 'Absorption' can also take place with certain X-ray energies due to X-ray diffraction. The atomic lattice of the absorbing metal acts as a diffraction grating, producing X-ray diffraction lines or spots, off the axis of the incident beam, thus effectively reducing the transmitted beam intensity.

The relative importance of these four absorption processes is indicated in Fig. 3.6, which is a plot of the linear absorption coefficients related to the

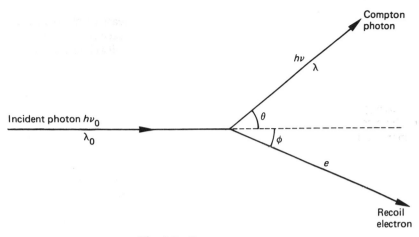

Fig. 3.5 Comptom scatter

different processes, together with the total linear absorption coefficient, μ, calculated for an iron absorber for different radiation energies. Figure 3.7 shows the total values of μ for three different metals.

It will be seen from Fig. 3.7 that for iron there is a broad region of minimum absorption at around 9 MeV, which corresponds to a maximum penetration. Relating this to an X-ray output curve such as Fig. 3.4, it means that the maximum penetration is obtained in steel with approximately 20 MV X-rays,

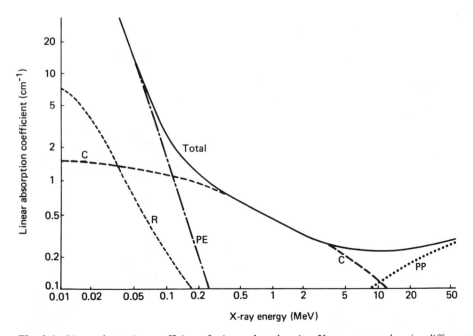

Fig. 3.6 Linear absorption coefficients for iron, plotted against X-ray energy, showing different components of the total scatter: C, Compton scatter; PE, photoelectric scatter; PP, pair production; R, Rayleigh scatter

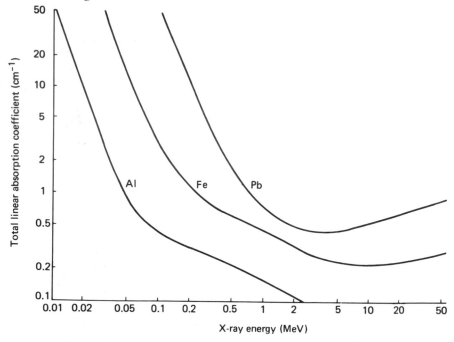

Fig. 3.7   Total linear absorption coefficients for lead, iron and aluminium

and that with higher energies no greater thickness penetration will be obtained.

3.3.2   Absorption curves

As stated earlier, linear absorption coefficients, μ, are defined in terms of narrow-beam absorption, whereas most radiography uses broad-beam conditions. A considerable amount of useful information can be obtained from comparable pairs of experimental absorption curves, for broad- and narrow-beam conditions, for a range of radiations. One experimental method of obtaining these is shown in Fig. 3.8: examples of pairs of absorption curves are shown in Figs 3.9 and 3.10 and derived data in Figs 3.11 and 3.12. In Fig. 3.8(a), a collimator prevents all scattered radiation from the absorber from reaching the detector, so that this set-up represents narrow-beam, true absorption conditions. In Fig. 3.8(b), there is a wide radiation field, i.e. broad-beam conditions. In Figs 3.9 and 3.10, the lower curve corresponds to narrow-beam conditions and the slope of the curve is directly related to the absorption coefficient. With X-rays (Fig. 3.9), it can be seen that the value of μ increases with absorber thickness as the spectral composition of the X-ray beam changes with absorption. With a cobalt-60 gamma-ray source, which is almost monoenergetic, the narrow-beam curve is virtually linear (Fig. 3.10). The separation between the two absorption curves is a measure of the build-up factor $(1 + I_S/I_D)$, and experimental values are shown in Figs 3.11 and 3.12. These values are likely to vary slightly with the precise geometric conditions of the absorption set-up and the spectral sensitivity of the detector, but there is

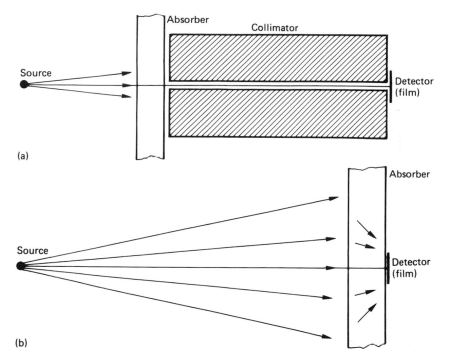

Fig. 3.8 Experimental set-up for measuring build-up factor

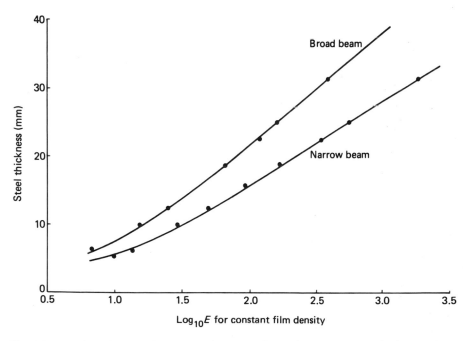

Fig. 3.9 Broad- and narrow-beam exposure curves for 200 kV X-rays, using lead intensifying screens (0.05 and 0.5 mm)

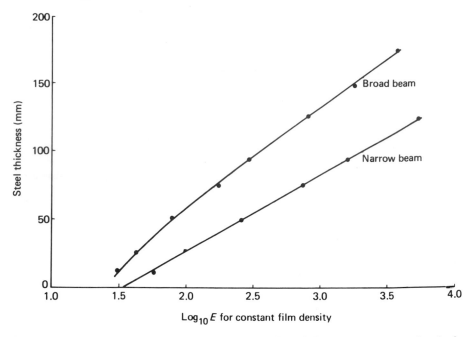

Fig. 3.10 Broad- and narrow-beam exposure curves for cobalt-60 gamma-rays, using lead intensifying screens (0.1 and 0.5 mm)

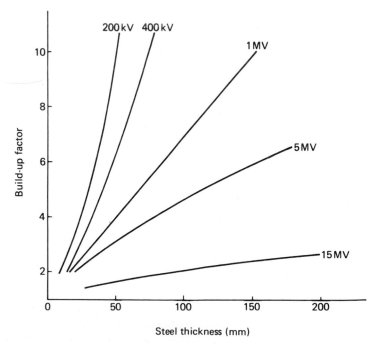

Fig. 3.11 Experimental values of build-up factor for different thicknesses of steel and different X-ray energies

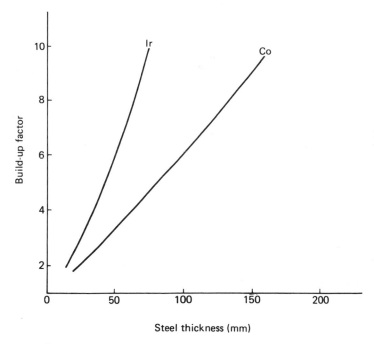

Fig. 3.12 Experimental values of build-up factor for different thicknesses of steel and different gamma-ray sources

close agreement between the results obtained by different investigators with different methods, and with the very limited number of published calculated values. The values in the figures are for uniform thickness plates; under 'poor' scatter conditions with an irregular-thickness absorber, they can be much higher. Values of $(1 + I_S/I_D)$ up to 40 have been reported.

It may be noticed that the broad-beam curve becomes almost straight at large thicknesses and this leads to the concept of an 'effective or equilibrium linear absorption coefficient' for broad-beam conditions, which corresponds to practical half-value and tenth-value thicknesses.

It is interesting to note that if this equilibrium value for gamma-ray sources is taken to calculate the corresponding equivalent X-ray kilovoltage values:

for iridium-192 gamma-rays: 0.72 MV X-rays,
for cobalt-60 gamma-rays: 2.3 MV X-rays,

i.e. both are effectively very-high-energy sources.

3.3.3 Contrast sensitivity

The effects of build-up factor and absorption coefficient values in radiography can be combined with the film parameters (see Section 3.6), and it is convenient to develop the equations at this stage, as it is fundamental to much of the later development of radiographic technique parameters.

Consider the radiography of a uniform thickness plate, thickness x, in which there is a small machined step of thickness Δx (Fig. 3.13). If Δx is small, it may be assumed that the values of I_S/I_D and μ do not change over the thickness range x to $(x + \Delta x)$. On a film (after processing) there is a density change from D to $(D - \Delta D)$ due to Δx, and over this small range of density the slope of the film characteristic curve (D against $\log_{10}E$—see Section 3.6) will be constant, G_D, so it is possible to write

$$D = G_D \log_{10} E + K \tag{3.9}$$

where $E = It$ is the exposure given to the film in time t and K is a constant. Then

$$\Delta D = 0.43 \, G_D \frac{\Delta I}{I} \tag{3.10}$$

the basic absorption equation is

$$I = I_0 \exp(-\mu x)$$

so $\Delta I = -\mu I \Delta x$ $\tag{3.11}$

Thus, if scattered radiation is neglected, by combination of equations 3.10 and 3.11,

$$\Delta x = -\frac{2.3 \Delta D}{\mu G_D} \tag{3.12}$$

This is an important case and will be referred to later, but the more important radiographic case is where scatter is present, and if it is assumed that scattered

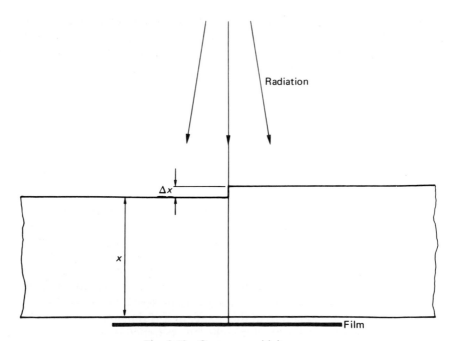

Fig. 3.13   Contrast sensitivity

radiation produces a uniform intensity I_S across the film and that the direct, image-forming radiation is I_D, it is necessary to put $I = I_S + I_D$ in equation 3.10, and $I = I_D$ in equation 3.11, giving

$$\Delta x = \frac{2.3 \Delta D}{\mu G_D} (1 + I_S/I_D) \tag{3.13}$$

If now it is assumed that ΔD is the smallest density-difference which the eye can discern on an illuminated film radiograph, Δx becomes the smallest thickness change which can be detected in a specimen of thickness x and the ratio $\Delta x/x$ is called the 'thickness sensitivity' or *contrast sensitivity*. This ratio is frequently given as a percentage value

$$S = \frac{\Delta x}{x} 100 = \frac{2.3 \Delta D [1 + I_S/I_D] 100}{\mu G_D x} \tag{3.14}$$

The negative sign can be omitted as it merely means that if Δx is positive, ΔD is negative.

It should be noted that μ is the linear absorption coefficient *measured under narrow-beam conditions* (equation 3.11), and that $(1 + I_S/I_D)$ is the build-up factor.

This is a fundamental equation in film radiography, as it relates the contrast parameters in a radiographic technique:

(1)   radiation parameters, $(1 + I_S/I_D)/\mu$,
(2)   film parameters, G_D,
(3)   viewing conditions, ΔD (minimum),
(4)   specimen thickness, Δx and x.

The important factor, so far as the radiation energy is concerned, is therefore not the build-up factor alone, but the ratio (build-up factor/μ), which rises to a maximum and then decreases, so that with high-energy X-rays the attainable thickness sensitivity begins to improve again.

3.3.4   Radiation intensity—units

Much of the preceding sections have been concerned with ionising radiation quality—wavelength, kV, MeV, etc. It is obviously necessary to consider also radiation quantity and intensity.

There are two systems of units in use, the newer SI system not yet having been widely adopted in industrial radiology.

In the older system, the quantity, or *exposure*, of ionising radiation is measured in *roentgens* (R), defined as the amount of radiation that produces ions carrying 1 esu of charge of either sign in $1\,cm^3$ of dry air at $0°C$ and $760\,mm\,Hg$ pressure. The roentgen is thus a measure of the energy absorbed by a given volume of air in a beam of radiation, and corresponds to $83.3\,ergs\,g^{-1}$. The output of radiation sources may be quoted in $R\,min^{-1}$ at a distance of 1 metre (often shortened to Rmm). The milliroentgen (mR) is widely used. As exposure is energy, in the SI system a separate radiation unit is not necessary and the coulomb kg^{-1} is used: $1\,C\,kg^{-1} = 3876\,R$.

A unit of *absorbed dose* is necessary and in the old system this is the *rad*, defined as the amount of any ionising radiation that results in the absorption of

100 ergs g^{-1} in any material, so that in air the rad and the roentgen differ only slightly in *numerical* values, and are virtually interchangeable for radiation energies less than 3 MeV. The roentgen, however, is a unit of exposure and the rad of absorbed dose and these are different concepts. The SI unit of absorbed dose is the *gray* (Gy), which is equal to 1 J kg^{-1}, or 100 rads.

The third unit in the old system is the *rem* (roentgen–equivalent–man). The corresponding SI unit is the *sievert* (Sv): 1 Sv = 100 rems. This is the equivalent dose used for protection purposes and is defined as 1 rem = 1 rad × (QF) where QF is the quality factor for a particular type of radiation. The QF is a means of relating absorbed doses of different radiations which have the same biological effect. For X-rays and gamma-rays QF = 1.

In practical X-radiography, the output of an X-ray set is most frequently quoted as the current across the X-ray tube, mesured in milliampères (mA), and a similar situation exists in gamma-radiography, where the strength of a radioactive source in curies is taken as a measure of radiation output. The *curie* (Ci) was originally defined as the disintegration rate of the quantity of radon in equilibrium with one gram of radium element and then redefined as the quantity of a radionuclide in which the number of disintegrations is 3.7×10^{10} per second. In the SI system, the curie is considered to be superfluous and a new unit, the *becquerel* (Bq) equal to a reciprocal-second, is used to specify radioactive source strengths. Thus

$$1 \text{ Ci} = 3.7 \times 10^{10} \text{ Bq}$$

As ionising radiation travels in straight lines outward from a source, i.e. it is not focussed, the simple inverse square law applies. Thus the radiation intensities at two points, distances d_1 and d_2 from a source, are related:

$$\frac{I_1}{I_2} = \frac{d_1^2}{d_2^2} \tag{3.15}$$

3.4   X-RAY SOURCES

As already stated, X-rays are generated when a beam of high-energy electrons is stopped suddenly by impingement on to a metal target, so that the essentials of an X-ray tube are (Fig. 3.14):

(1)   a filament, as a source of electrons—the cathode,
(2)   a target—the anode,
(3)   a high-voltage supply which can be connected across the cathode and anode.
(4)   a means of supporting the anode and cathode in a very high vacuum space—the actual X-ray tube envelope.

In older tubes, the tube envelope was glass and the vacuum could be obtained either by operating the tube on vacuum pumps, or by sealing off the tube after evacuation. Special non-filament tubes are also available for some applications (see Section 3.11.5). The filament is located in a recess in the cathode, called the focussing cup, which helps to produce a narrow, well-defined beam of electrons on to the anode. In some newer tubes, the electron

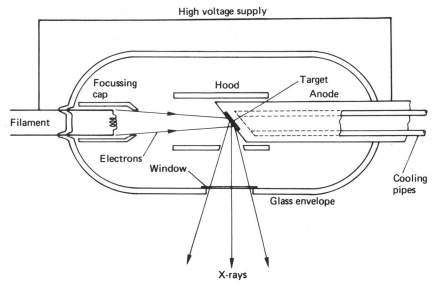

High voltage supply

Focussing cap

Hood

Target
Anode

Filament

Electrons

Window

Cooling pipes

Glass envelope

X-rays

Fig. 3.14 Conventional design of glass envelope X-ray tube

beam is focussed with electron lenses, to produce 'microfocus X-ray tubes'. In many modern tubes, particularly those for operation in the 200–400 kV range, the glass envelope is replaced by a metal envelope with ceramic disc insulators at each end to hold the anode and cathode assemblies and provide the necessary electrical insulation. This leads to a smaller, more robust X-ray tube than a glass envelope design.

At the anode, only a small proportion (1–10%) of the energy of the electrons is converted to X-rays and most becomes heat energy. The area of impact of the electrons on the anode therefore is usually a tungsten disc, which has a very high melting point; this is bonded into the copper anode, which may be air-, water-, or oil-cooled. The anode may also be hooded (Fig. 3.14) to restrict the X-ray beam, and to prevent charge build-up on the tube envelope from scattered electrons.

The anode face is angled (Fig. 3.15) so that with a horizontal beam of electrons as shown, the beam of X-rays spreads out from a small, elongated area on the target, but as seen from the centre of the X-ray field, the *effective* width of the source of the X-rays is considerably smaller than the area of the target on which the electrons are incident.

Typical effective source widths—usually called 'focal-spot sizes' or 'focus sizes'—are between 1 and 5 mm, whereas the actual area on which the electrons impinge may be several times larger. If the electron beam from the filament is made rectangular in cross-section, it is possible to have an approximately square effective focal-spot size, say 1×1 mm or 4×4 mm. It is not usually possible to have a uniformly emitting focal spot, or a completely regular shape, and the effective size of the focal spot cannot be measured or specified with much accuracy. The effective size of the focal spot is also slightly different at the edge of the X-ray field, e.g. smaller at A, larger at B in Fig. 3.15.

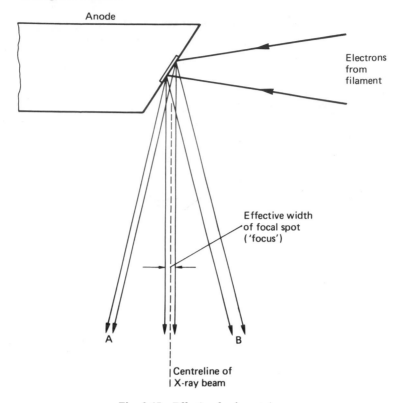

Fig. 3.15   Effective focal-spot size

The area of the anode on which the electrons impinge has to dissipate all the heat which is generated, so that small focal-spot tubes tend to have lower X-ray outputs because of the difficulties in arranging sufficient cooling of the anode.

There are a number of other anode configurations, the most important being shown in Fig. 3.16. With the configurations shown in Fig. 3.16(a) and (b), a panoramic X-ray beam is obtained, and it is also possible to design the tube in the form of a long, thin anode which can be inserted into pipes (see Section 3.10.1). The configuration in Fig. 3.16(c) is a transmission target, often used with high-energy machines. The X-rays are emitted in the forward direction.

There are two basic designs for X-ray sets.

(1)   The X-ray tube in the protective shield is connected to the high-voltage generators by one or two HT cables. This gives a relatively small, lightweight tube head for easy manoeuvring into position for radiography. A modern 420 kV tube head of this type need weigh only around 100 kg.
(2)   The X-ray tube and HT generators are built into one tank, with oil or gas insulation. This eliminates HT cables which have often been a cause of unreliability. Tank-type sets range from about 60–300 kV.

In the first design, it is common practice for the higher-energy sets to have a centre-earthed tube, with, for example, −150 kV on one end and +150 kV on the other, giving a total of 300 kV. Alternatively, one end of the tube can be earthed: if this is the anode end, forced liquid cooling is easy to arrange and does not require an insulated pumping system. Often the cooling liquid is an insulating oil which serves the double purpose of insulation and cooling.

Such conventional X-ray tubes operate up to about 420 kV, have focal-spot sizes in the range 0.5–4.0 mm diameter and produce tube currents from 1 to 25 mA, depending on the focal-spot size and the cooling system. Dual-focus tubes with two filaments are common.

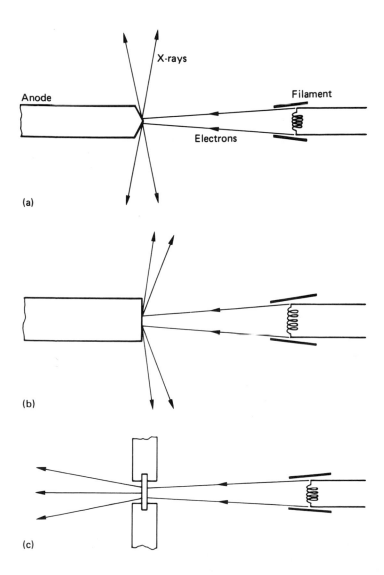

Fig. 3.16 X-ray tube anode configurations: (a) conical anode, giving a radial panoramic beam, (b) flat anode, giving a backwards panoramic beam, and (c) transmission target for forward beam

In recent years, microfocus X-ray tubes have been developed with focal-spot sizes from 0.2 mm down to around 5 μm. These focal-spot sizes are obtained by focussing the electron beam from a special design of filament on to the target. Very small focal-spot sizes give rise to a danger of overheating and pitting on the target and some of these tubes are built in demountable form, so that the target and filament can be easily replaced. Several microfocus X-ray tubes are available to operate up to 150 kV and some up to 360 kV. They have special radiographic techniques and applications (see Section 3.9.2). The accurate measurement of focal spots in the 5–20 μm range is difficult. One method which has been proposed (British Standard BS:6932:1988) is to image a small tungsten ball at a large magnification and calculate the source diameter from the geometric unsharpness in the image. For accuracy this method requires a microdensitometer to produce a density trace across the image, but a micro-scale can produce a useful result with moderate accuracy. The X-ray output obviously depends on the focal-spot size and on the voltage: values of

90 kV   0.5 mA   15 μm diameter focus
(45 watts anode power)

seem to be typical of what has been achieved, to date, for continuous operation.

Rotating anode X-ray tubes, in which the electrons impinge on to the periphery of a rotating tungsten disc, are widely used in medical radiography, with very short exposure-times, and small focal spots, but for the longer exposures usually needed for industrial work they are less advantageous and are not physically very robust. 'Flash X-ray tubes' are very special in design and their application and construction will be briefly described in Section 3.11.5.

For voltages up to about 400 kV, the commonest method of high-voltage generation until a few years ago was a step-up transformer with a rectifier. There were several circuits, e.g. Kersley, Villard, Graetz, and Greinacher. Strictly, a rectifier is not essential, as the X-ray tube can act as its own rectifier, but this shortens the life of the tube, and also means that the X-rays are emitted only over half the voltage cycle. Most of the circuits used are designed to obtain some output over the whole cycle, and by using a capacitor (or capacitors) to increase the effective voltage applied to the tube, smaller transformers are possible.

In recent years, high-frequency, HT generators have been developed, employing thyristor invertors and Cockroft–Walton multiplier circuits. These have enabled smaller power generators to be built, with more stable characteristics and more accurate control of both voltage and current. The outputs can be constant to better than 0.1% with a total voltage ripple of less than 1.5% peak-to-peak.

On some of the earlier circuits, the X-ray output (mA) was measured indirectly and values were not particularly accurate. In this energy range, up to 420 kV, most X-ray sets can operate at any chosen voltage from about 60 kV up to the design limit. The lower-energy tubes often have a beryllium window, so that very-low-energy X-rays are not heavily absorbed in the tube wall.

3.4.1   High-energy X-ray sources

Above about 450 kV, the conventional design of X-ray tube and HT generator becomes large and cumbersome, and while a few X-ray sets of this type have been built for 600 kV (maximum), there are now no commercial X-ray sets operating between 420 and 1000 kV; for 1000 kV and higher, special designs of X-ray set have been devised.

Tuned-transformer (resonance transformer)

This type of set, first developed in 1940 for 1 MV and 2 MV by Charlton and Westendorp, is no longer marketed, but some sets still exist and some have been operating for over forty years.

In this design, a one-end-earthed air-core transformer has a multisection X-ray tube along its axis instead of an iron core, the whole assembly being in a gas-insulated tank. The transformer can be resonated at a multiple of the input frequency (180 Hz) and is connected to tube segments so as to allow electron acceleration at a series of stages up to the full operating voltage. The 1 MV set had a glass tube 160 cm long with twelve segments.

Van de Graaff generators

These also use a sectionalised tube, with an earthed anode and a high-resistance potential divider. An electrostatic charge is sprayed on to a high-speed travelling belt, by which it is carried to the filament end of the tube, where it gives up the charge to a high-tension hemisphere. The charge builds up to the limit set by the insulation, but as the X-ray tube provides a steady load, an equilibrium condition can be set up when there is a current along the tube. Van de Graaff sets have been built to operate between 1 and 8 MeV with tube currents between 0.25 and 1 mA. A few are used for industrial radiography, but their main market has been for irradiation work. They can have a very small focal-spot size (0.1 mm).

Betatrons

These machines accelerate electrons in a circular orbit by using an alternating magnetic field. The electrons travel in a toroidal vacuum chamber which is placed between the poles of a powerful electromagnet. The electrons travel round a circular path in a constant orbit, increasing in energy as the magnetic flux is increased: at maximum energy in each pulse they are deflected on to a target to produce X-rays. In industrial betatrons, the electrons are accelerated to an energy in the range 15–31 MeV. These very high energies are necessary to obtain a reasonable output. A very small focal-spot size (<0.5 mm) is possible, but X-ray output is comparatively low and the machines are large and heavy. A few industrial betatrons are still in use, but generally have been replaced by linacs (see below).

In the USSR, a number of small, portable betatrons have been built (2–6 MeV) and are reported to be widely used, but they have an extremely low X-ray output (e.g. 4 MeV, 1 R min^{-1} at 1 m), which would appear to put severe limitations on their applications.

Linear electron accelerators (linacs)

For the production of X-ray energies above 1 MeV the linac appears to be the most useful modern X-ray machine, and linacs have been built to operate up to 25 MeV. In the linac, electrons from a filament gun are introduced at a low voltage into the cavity of a linear waveguide carrying either a travelling or standing radiofrequency wave, generated by a magnetron or a klystron. The electrons increase in energy from the radiofrequency wave, by arranging for the velocity of the electrons and the phase velocity of the microwaves to be equal, so that the electrons stay in the same high-frequency phase. After acceleration down the centreline of the corrugated waveguide, the electrons are focussed on to a target to produce X-rays, usually in the transmission mode (Fig. 3.17). Most industrial linacs use the 10 cm wavelength, but a few use 3 cm microwaves. Typical performances are:

$$2 \text{ MeV}: 2 \times 2 \text{ mm focal spot, } 200 \text{ R min}^{-1} \text{ at 1 m}$$

$$8 \text{ MeV}: 3 \times 3 \text{ mm focal spot, } 2000 \text{ R min}^{-1} \text{ at 1 m}$$

$$12 \text{ MeV}: 1 \times 1 \text{ mm focal spot, } 6000 \text{ R min}^{-1} \text{ at 1 m}$$

These machines therefore have a very large output of X-rays: they are fairly large and heavy (2 MeV, 800 kg) with the main waveguide tube about 1 m long, but have proved to be very reliable. There are also a few small linacs specially designed to be lightweight and portable for site-work applications. A typical specification for these is: 3.5 MeV; 100 R min^{-1} at 1 m; 2 mm focal spot; weight of 'head', 40 kg. In some of these machines the head is separated from the generator by a flexible waveguide and much of the weight of the head is taken up by a collimator to restrict the X-ray beam and minimise extraneous scattered radiation.

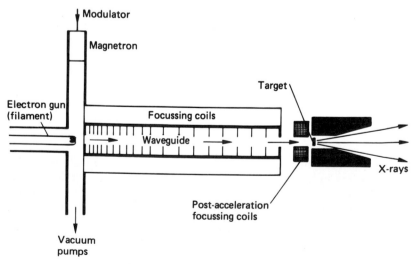

Fig. 3.17   Linear electron accelerator (linac)

Microtrons

These form another possibility for generating high-energy X-rays, but very few microtrons are at present in industrial use. The microtron is a form of circular–orbit accelerator in which the electrons are accelerated in a microwave resonator which is placed in a homogeneous magnetic field. The electrons have different radius orbits with a common tangent in the resonator. Klystron-operated 8 MeV microtrons have been built which are claimed to have outputs of 6000 R min^{-1} at 1 m which is higher than that of a comparable energy linac, but microtrons have a reputation for being difficult to operate and maintain.

3.4.2 Comparison data

Table 3.1 gives comparison data on the characteristics of typical X-ray equipment.

With voltages up to 420 kV, the effective focal-spot size can be measured with a simple pinhole—a very small hole in a lead sheet is placed midway between the X-ray tube and a film. The pinhole 'images' the focal spot on the film and, provided the pinhole is less than one-quarter of the focal-spot size, adequate accuracy of focal-spot size measurement can be achieved.

With high-energy X-rays, the pinhole method does not work because the image obtained is of too low contrast. If a sharp metal edge of appropriate thickness is imaged, accurately aligned with the radiation beam, the unsharpness of the edge image is a combination of film and geometric unsharpness. If the geometric unsharpness is made large, the source diameter can be calculated from its value, measured with a microdensitometer.

A more accurate method, but one requiring much more elaborate equipment, is to scan the radiation source with a crossed slit forming a pinhole, with a radiation detector behind. An accurate scanning frame and a stepping motor is necessary. In practice with linacs the X-ray target is physically limited, so the physical dimension automatically gives a maximum value to the focal spot size.

Table 3.1 X-ray equipment characteristics

Type	Maximum voltage (kV)	Focal-spot size (mm)	X-ray output R min^{-1} at 1 m	Weight of head (kg)
Tube head/cables	120	0.8 × 0.8	— (10 mA)	9
	420	3.5 × 3.5	50 (10 mA)	100
Minifocus	160	0.4 × 0.4	— (4 mA)	3.5
Microfocus	150	0.02 × 0.02	— (0.5 mA)	50
Tank type	200	3 × 3	— (6 mA)	100
	300	3 × 3	16 (6 mA)	120
Resonance	1000	7 × 7	50 (3 mA)	1350
Linac	1000	2 × 2	50	1800
Van de Graaff	2000	1 × 1	70	1850
Minac	6000	2 × 2	300	40
Linac	8000	2 × 2	3000	1850
	15000	3 × 3	6000	2000
Betatron	24000	0.2 × 0.2	180	3000

3.4.3   X-ray field size

With X-ray tubes up to 1000 kV, the field size of the X-ray beam is large and depends largely on the shape of the target. There is a slight variation in X-ray intensity across the field (about 5%), sometimes called the 'heel-effect', but this is too small to be much of a practical problem in normal radiography.

With megavoltage X-rays, however, the physics of X-ray production means that if the electron beam is incident normally on to a transmission target, the X-ray field becomes progressively narrower in width as the electron energy increases. At 15 MeV the half-width to half-intensity is only about 6°, so that to cover a large-area specimen at one exposure, extra-long focus-to-film distances may be necessary, which means longer exposure-times.

There are two solutions. One is to use a field–flattening filter, close to the X-ray target. Such a filter is thicker at the centre and thinner at the edges and it 'flattens' the field by absorption. This of course means that the effective output of the X-ray machine is reduced. The second method is applicable only to linacs. When the electrons have been accelerated down the main waveguide, they are refocussed on to the target: if this is done with short-focus electron lenses, some of the electrons strike the target at a considerable angle to the normal and this has the effect of substantially widening the resulting X-ray field. With a 5 MeV linac, the half-angle to half-intensity was increased from 11° to 27° by using this method of post-acceleration focussing.

3.5   GAMMA-RAY SOURCES

There are a few naturally-occurring radioactive materials and several hundred artificial radioisotopes.

Radioisotopes emit, α-, β-, or γ-rays, or a combination of these, and a few emit neutrons. Of those emitting gamma-rays, only a very few have properties which make them suitable for industrial radiographic applications.

A radioactive atom disintegrates by several processes which occur irregularly with time, but when taken statistically over a long period, a general law can be stated:

$$\Delta N = \lambda N \Delta t$$

where there are N nuclei, and ΔN nuclei disintegrate in time Δt. The disintegration constant, λ, is usually converted into a half-life, $T_{1/2}$, where

$$T_{1/2} = 0.693/\lambda$$

and $\qquad I = I_0 \exp[(-0.693/T_{1/2})t]$ $\hfill$ (3.16)

Thus, for example, after two half-lives, a source has decayed to one-quarter of its original strength.

The strength of a source is measured in curies [or in the SI system, in becquerels, (Bq)]. See Section 3.3.4. The *specific activity* of a source, usually measured in curies per gram (Ci g^{-1}), and the *specific gamma-ray emission* (sometimes called the *K-factor*), measured in roentgens per hour at 1 cm from a 1 mC source, are important properties of gamma-emitting radioisotopes.

Until about 1955, natural radioactive sources, usually radium and occasionally mesothorium, were used for gamma-radiography; in the UK, radon, a gas in the radium series, was extracted from radium in solution, and provided a very small diameter high-strength radioactive source, but with a very short half-life (3.8 days). The radon was absorbed on to charcoal to give a source of less than 0.5 mm diameter and initial strengths of 10–20 curies were possible. The radon-extraction process, however, was hazardous; radium is exceedingly expensive as well as hazardous (half-life 1590 years), and none of these sources is now produced for industrial radiography.

Almost all gamma-radiography sources used today are cobalt-60, iridium-192, or ytterbium-169, with a very small usage of caesium-137, thulium-170, or sodium-24. A few other radioisotope sources are used in thickness gauging. It should be noted that the number following the element name is the atomic mass: thus

$$\text{cobalt-60} \qquad ^{60}\text{Co} \qquad \text{Co-60} \qquad ^{60}_{27}\text{Co}$$

are all representations of a cobalt atom of atomic number 27 and a total of 60 protons plus neutrons in the nucleus.

Artificial radioisotopes are produced in atomic piles, either by neutron irradiation or as fission products which are extracted from used fuel elements. The requirements for a suitable gamma-ray source for industrial radiography are:

(1)   a reasonable half-life—reactivation or frequent exchange can be costly,
(2)   a high specific activity so as to obtain a high output of radiation from a small diameter source,
(3)   a high K-factor,
(4)   a reasonably-priced basic material.

It would be desirable to have available a range of gamma-ray sources, from the equivalent of (say) 50 kV X-rays up to 8 MV X-rays, so as to be able to produce gamma-radiographs as good as X-radiographs on any specimen thickness and material, but such a choice does not exist at present.

3.5.1   Cobalt-60

These are made by neutron activation of the natural stable isotope, cobalt-59, which is a hard metal of density 8.9 g cm^{-3}. It decays by the emission of a 0.31 MeV electron and then two gamma-rays of energies 1.17 and 1.33 MeV are emitted. The gamma-photons are in cascade so they are emitted at the same rates. The half-life is 5.3 years and the radiation output is 1.3 Rhm Ci^{-1}. Having a long half-life, cobalt sources also take a long time to activate. Most gamma-radiography cobalt-60 sources are of activity 20–30 Ci (740–1010 GBq), but a few fixed installations with 1000–2000 Ci sources are in use. At present, a 20 Ci source is likely to be 3×3 mm in size, or larger. As a result of the high-energy gamma-ray spectrum lines, cobalt-60 sources are suitable only for the radiography of thick steel specimens and some codes of good practice limit their use to steel thicknesses greater than 50 mm.

3.5.2   Iridium-192

There are two naturally-occurring isotopes of iridium, Ir-191 (38%) and Ir-192 (62%). When the natural metal is subjected to neutron irradiation, two radioisotopes are produced, but as Ir-194 has a half-life of only 19 h, after a few days the resulting source consists almost entirely of Ir-192, which has a half-life of 74 days. Iridium is a very hard metal of density 22.4 g cm^{-3}. The radiation output is 0.5 Rhm Ci^{-1}. The decay of an iridium-192 source is complex, and the gamma-ray spectrum contains a large number of spectrum lines of different intensities (see Table 3.2).

A wide range of source sizes (0.5 × 0.5 mm to 4 × 4 mm) is used in industrial radiography, with source strengths up to 100 curies (3720 GBq). Iridium sources are used for the examination of steel specimens from about 10 to 100 mm thick.

Table 3.2   Gamma-ray spectrum of Ir-192: principal lines

Energy (MeV)	Relative intensity (%)
0.200	0.5
0.210	4.1
0.280	0.4
0.296	35.8
0.308	37.1
0.316	100.0
0.374	0.8
0.416	0.7
0.468	56.8
0.484	3.5
0.489	0.4
0.589	5.3
0.604	9.9
0.612	6.4
0.884	0.4
1.062	0.07

3.5.3   Caesium-137

This is a common fission product which can be extracted chemically from spent fuel-rods. Its main importance is in the rather long half-life of 30 years. It decays by beta-emission to an isomeric form of Ba-137, from which a single gamma-ray spectrum line of energy 0.66 MeV is emitted. The radiation output is 0.37 Rhm Ci^{-1}, and it cannot be made into such high-intensity, small diameter sources as is possible with iridium-192. Caesium-137 sources were originally produced from encapsulated powder, but today they are formed into a glass-like solid, which is inherently safer in the case of accidents.

3.5.4   Thulium-170

The need for a low-energy gamma-ray source has led to the production of Tm-170 sources from naturally-occurring Tm-169, even though thulium is a very expensive rare-earth element. The half-life is 127 days and the radiation output intensity is very low; the actual output value depends on the physical size of the source as there is marked self-absorption. Values around 0.003 Rhm Ci^{-1} have been quoted.

The decay is complex: two beta-particles of energy 0.968 and 0.884 MeV are emitted; the nucleus is then stabilised by the emission of a 0.084 MeV gamma-ray and by internal conversion, resulting in a second gamma-ray of energy 0.052 MeV, but only about 8% of the disintegrations result in the gamma-radiation. The high-energy beta-particles (electrons) are internally absorbed to produce a 'white' spectrum of bremsstrahlung with the peak energy at around 0.9 MeV. The radiographic results are therefore also complex. With a physically-small Tm-170 source and a very thin specimen, the radiographic image is in part, at least, produced by the two low-energy gamma-ray spectrum lines. With larger sources and thicker specimens, the gamma-ray lines are absorbed and lost, and the radiograph is produced largely by the high-energy bremsstrahlung, which results in a low-contrast, poor-quality image.

3.5.5 Ytterbium-169

In recent years, Yb-169 sources have been marketed in Europe for thin-section radiography.

Ytterbium is a very scarce and very expensive rare-earth element, and Yb-169 sources are produced by neutron irradiation of a mixture of stable isotopes, allowing a few days for the short-lived products to decay. Ytterbium-169 has a half-life of 31 days, an output of 0.125 Rhm Ci^{-1}, and a complex spectrum (see Table 3.3).

Recently, in spite of the high cost, Yb-169 sources have been made from an artificially enriched natural ytterbium, to produce higher source strengths.

Table 3.3 Gamma-ray spectrum of Yb-169: principal lines

Energy (keV)	Relative intensity (%)
63.1	45.0
109.8	18.0
130.5	11.5
177.2	21.0
198.0	33.0
308.0	10.0

The principal application at present is for the radiography of welds in thin-wall tubing, for which very small diameter sources are desirable (e.g. 0.3 – 0.6 mm diameter sources). Source strengths of around 1.0 Ci are now available in a 0.6 mm spherical source. The availability of gamma-rays with radiographic characteristics similar to those of a 150–250 kV X-ray source, makes Yb-169 a very important new source.

3.5.6 Other sources

Many other gamma-ray sources have been investigated for industrial radiography. There have been reports of europium-152, -154 being used in the USSR, and some interest in iodine-24 and xenon-133 in the USA. A valuable source, if it can be taken to the radiography site very quickly, is sodium-24. This has two very-high-energy gamma-ray spectrum lines (1.37 and 2.76 MeV), so it is suitable for very thick specimens, but its half-life is only 15 hours. There have also been investigations of the use of bremsstrahlung from beta-emitting sources, for radiography.

3.5.7   Gamma-ray source-handling equipment

Pellets of gamma-ray emitting materials are produced in standard sizes (0.5×0.5 mm, 1×1 mm, 2×2 mm, 3×3 mm, etc.) by the atomic energy authorities in various countries in special atomic reactors, and are sealed into standard patterns of source capsule. Once the material is activated, the radiation output depends only on the initial strength and the half-life: gamma-ray sources cannot be switched off. These sealed sources therefore cannot be handled with impunity, and there are three basic methods of providing protection for personnel:

(1)   introduction of absorbing material around the source capsule,
(2)   use of distance—the inverse square law applies,
(3)   use of time—rapid handling.

At one time, gamma-ray sources were held in storage or transport containers and transferred to an 'exposure container' for use, but today the usual practice is to keep the source in one container and add extra thicknesses of absorber if this is necessary for transport. There are four basic designs of exposure container:

(1)   the sphere with a removable cap (Fig. 3.18),
(2)   the rotating core container (Fig. 3.19),
(3)   the sphere from which the source is removed by a rod (Fig. 3.20),
(4)   the container in which the source is held and moved on a flexible cable (Fig. 3.21).

Type (1) containers are suitable only for low-intensity sources. With these source containers, the radiographic set-up should be arranged with an empty or a dummy container and the loaded container substituted as the final pre-exposure operation, so as to minimise handling-time. The cap of the container should be removable with a handle, as otherwise the operator's hand may be only a few centimetres from the source while the cap is removed.

In order to have the advantage of a lightweight, easily portable container, type (1) containers are sometimes designed for only a very limited handling-time. Extra care is therefore necessary in planning the setting-up procedure, to ensure that the limited handling-time is not exceeded.

Type (2) is a derivative of type (1) in which the removable cap has been eliminated. It can be opened from a distance and handled with a long handle.

Type (3) was built originally for pipeline work: the source, with some protection on three sides, is lifted out of a parent container, into a holder fixed to the pipe in the exposure position.

Type (4) is more versatile, intrinsically safer, and more suitable for larger sources in the 20–200 Ci strength range. In the closed position (Fig. 3.21), the source is at the centre of a large absorber: this is the safe storage position. For radiography, the collimator is put in the position required for radiography, and then the source on a flexible cable is moved along the connecting tube to the

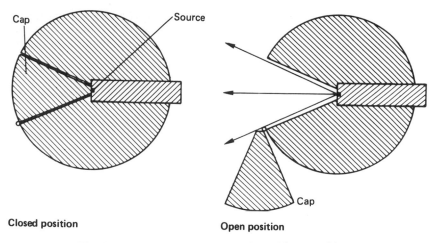

Fig. 3.18 Gamma-ray source container with removable cap

exposing position (the open position in Fig. 3.21). The flexible tube may be several metres long and the movement controller can be a similar distance behind the storage container, so that the operator is never less than (say) 20 m from the source during the short time when the source is fully exposed. The source movement may be manually operated, or electrically operated, or operated by compressed air. If electrically operated, timing clocks and delay mechanisms can be incorporated. For use with large cobalt-60 sources, the container in type (4) equipment may weigh 100–300 kg and can be on a trolley or small trailer; for use with ytterbium-169 sources, a similar design need weigh only 2–5 kg. Because of the relatively low energy of gamma-rays from Yb-169, small containers can be designed to clamp on to pipework. These carry a source on one side of a pipe, with a small film on the opposite side,

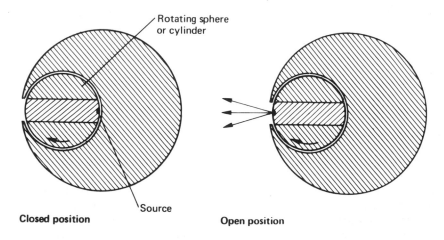

Fig. 3.19 Gamma-ray source container with rotating core

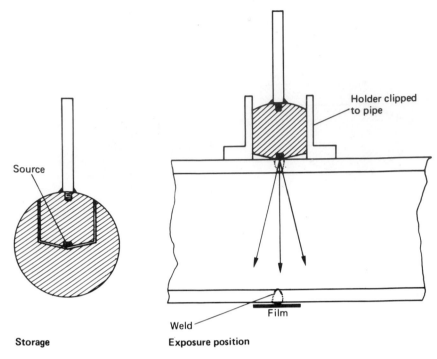

Fig. 3.20   Gamma-ray source container with detachable source, for pipe weld radiography

with sufficient radiation protection to allow exposures to be made without having to move personnel more than 1–2 metres away from the exposure site.

There are many variations on these basic designs of gamma-ray exposure container, and ancillary features, such as source position indicator lamps and overload prevention clutches, and general design quality are extremely important.

The material of these source containers obviously needs to be a high-density, radiation-absorbing metal, and nearly all early containers were made of lead. A sintered tungsten material (tungsten particles in a copper matrix) which has a much higher density (16–18 g cm^{-3}) has also been used: it is easily machinable but rather expensive. National and international transport safety regulations for gamma-ray source containers often contain extremely rigorous requirements regarding fire hazards, and do not allow the use of lead containers, so that modern designs use a combination of depleted-uranium cast into a thin containment shell of steel or brass, with some smaller tungsten-alloy components.

The obvious use for gamma-ray sources is for site work, and here the transportability of the source and its container are of great importance. There are strict regulations in most countries on the permissible external dose-rate on containers which are taken on to public roads, or moved by rail, or by air.

Gamma-ray sources are also used in fixed installations, in preference to large X-ray sets: they are intrinsically cheaper in capital cost than high-energy X-ray machines such as linacs, and with panoramic exposures or carousels it is

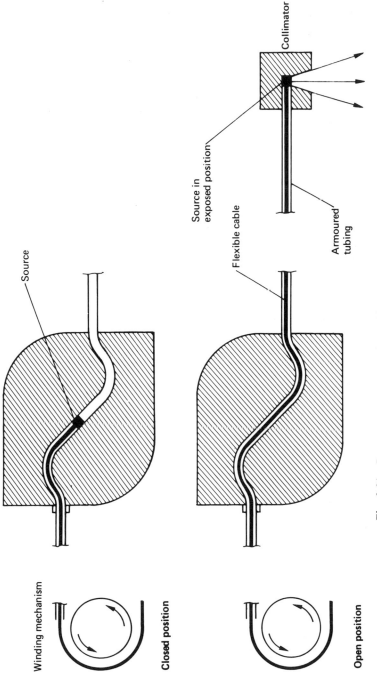

Fig. 3.21   Gamma-ray source container with source on flexible cable

possible in some applications to minimise the effects of the longer exposure-times needed by radiographing a large number of specimens during each exposure-time.

3.6   RECORDING OF RADIATION

When a beam of X-rays is incident on a specimen, the transmitted beam contains spatial variations in intensity which represent the X-ray image of the inhomogeneities in the specimen. X-rays are not directly perceptible by the human senses, so that this intensity distribution must be converted to some other variable, and for many years the most widely used converter has been a sheet of radiographic film. However, there are alternative detectors which will be discussed later in this section.

3.6.1   Radiographic film

A normal radiographic film consists of two emulsions of silver halide crystals in gelatine, coated on either side of a flexible film base, nowadays a polyester, with an adhesive substrate, anti-scratch layers, etc. Modern radiographic film is marketed in flat sheets of all the usual photographic sizes up to 30×43 cm, and in long lengths of strip film for wrapping round pipes. It may be loose-interleaved in boxes, envelope-packed for direct exposure, or have built-in lead intensifying screens in vacuum-sealed envelopes. There is an enormous literature on the photographic process and it is sufficient here to state that after exposure the film is processed (i.e. developed, rinsed, fixed, washed and dried), when it will be found that the X-ray intensities are represented by different degrees of film blackening—photographic densities—which can be seen by placing the processed film on an illuminated screen.

Photographic density, D, is defined as

$$D = \log_{10} I_o / I_t \tag{3.17}$$

where I_o is the intensity of light incident on one side of the film and I_t is the intensity transmitted through the film.

For a formal definition, the geometric conditions should be defined, but this is not necessary here. Thus a film which transmits one-hundredth of the incident light has a density of 2.0. For present purposes, the exposure given to a film, E, will be taken as the direct product of X-ray intensity, I, and exposure time, t.

$$E = It$$

If a number of small areas on a film are given different exposures, the film then uniformly processed and the densities measured, a curve can be plotted of density, D, against exposure, E (Fig. 3.22). More usually, the *characteristic curve* of D against $\log_{10} E$ is plotted (Fig. 3.23).

The characteristics of the latter curve for films exposed to X-rays are:

(1)   a minimum film density, a, which is not zero—the *fog level*—usually between 0.2 and 0.3,

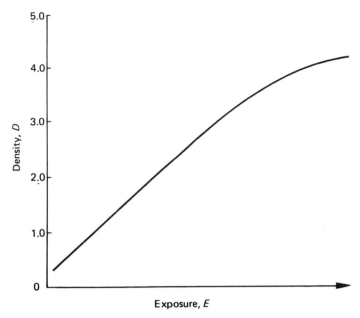

Fig. 3.22   Radiographic direct-type film: density/exposure curve

(2)   the toe of the curve (region A–B), in which the slope of the curve is increasing,

(3)   a region (B–C–D) in which the slope of the curve is more nearly constant,

(4)   a region above D where the curve begins to turn over to a maximum value, D_{max}: for most radiographic films, D_{max} is between 10 and 15.

A mathematical description of the $D/\log_{10}E$ curve requires a third or fifth degree polynomial, so most data is derived graphically.

The early work on photographic sensitometry was carried out with light and the region B–C was called the 'straight-line' portion of the characteristic curve; with X-rays it is not straight.

The slope of the D against $\log_{10}E$ curve at any point is called the *film gradient*, or *film contrast*, G_D, and is given by

$$G_D = \tan\phi = \frac{dD}{d(\log_{10}E)} \quad \text{(at density D)} \quad (3.18)$$

The term 'gamma, γ', used in light photography as the slope of the straight-line portion of the characteristic curve, has no real meaning on film exposed to X-rays, but average values of G between different densities are sometimes quoted as gamma-values or average gradients.

Derived curves of G_D against D are important in radiography and an example of these is shown in Fig. 3.24. It will be recalled that this parameter G_D was used in the fundamental contrast sensitivity equations given earlier (see equations 3.13 and 3.14).

One other important parameter can be obtained from a film characteristic curve. This is a *film speed number*. If the exposure, E, is measured in some

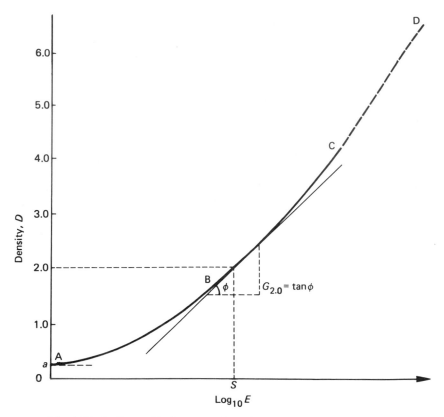

Fig. 3.23   Radiographic direct-type film: density/$\log_{10}E$ characteristic curve

standard units of radiation dose, the exposure, S (Fig. 3.23) required to produce an agreed arbitrary standard film density is a measure of film speed. The major film manufacturers market three, four, five, or even more, different radiographic films of different speeds, covering a speed range of at least 20:1, which are usually described in qualitative terms such as 'very-fine-grain', 'fine-grain', 'medium-speed', or 'high-speed'. There is no internationally agreed method of specifying either film speeds or film graininess. It is generally assumed that a slower, finer-grained film is capable of showing more, or better, image detail than a faster film, other things being equal. They also have a higher contrast (see Fig. 3.24). The measurement of film graininess (the visual impression) or granularity (instrument measurements) is complex. If a small aperture is scanned over a film of uniform density, the local fluctuations in density can be recorded, and the root-mean-square density fluctuation (σ_D) can be determined. It has been claimed that if a scanning aperture of 100 micrometres diameter is used, the values of σ_D obtained on radiographic film correlate closely with graininess values, and values of (G/σ_D) have been proposed as a 'film quality index' measurement. It is claimed that G/σ_D is a measure of (signal/noise) ratio.

The relationship between film graininess and image resolution is complex. When an X-ray quantum is absorbed in the film emulsion, it sensitises a silver

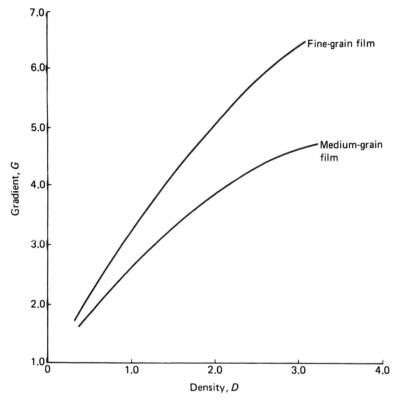

Fig. 3.24 Radiographic direct-type film: gradient (contrast)/density curves

halide crystal; furthermore, it may have sufficient energy to release electrons which can reach and sensitise adjacent silver halide crystals. One X-ray quantum can therefore sensitise a small volume of silver halide crystals, thus producing a small disc instead of a point image and this is equivalent to producing an *inherent* or *film unsharpness*, U_f. If, in Fig. 3.25, a sharp metal edge is laid on the film, and the image of this edge on the processed film is examined, it will be found that an X-ray quantum absorbed in a silver halide crystal just to the left of O has sensitised silver halide grains to the right of O, and the density distribution across the image follows curve B in practice, rather than the idealised curve A. The width of the unsharpness, U_f, can be measured in mm, analogously to geometric unsharpness. Obviously, high-energy quanta will produce a greater effect, and measured values of U_f are given in Fig. 3.26. Analogous values for gamma-ray sources are given in Table 3.4.

Film unsharpness depends more on the silver/gelatine ratio in the film emulsion, than on the actual film-grain size, and experimentally appears to be almost independent of film graininess or film granularity in its magnitude. Film graininess, measured in terms of the blending distance at which no graininess can be discerned, was found to increase by only 40% between 100 kV X-rays and 10 MV X-rays for a particular film, whereas U_f increased

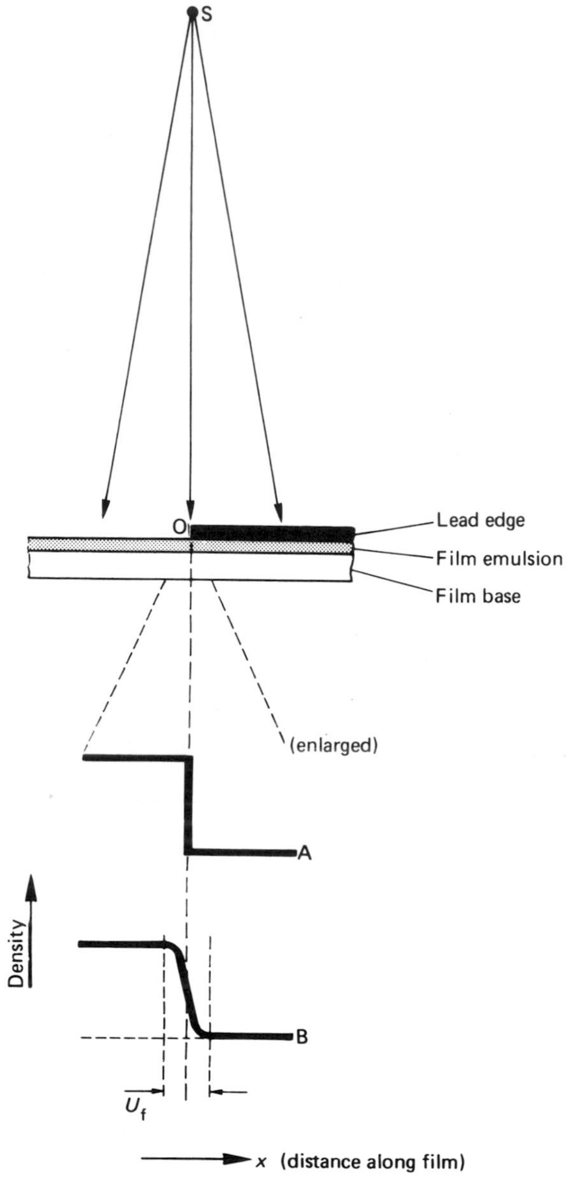

Fig. 3.25   Film unsharpness, U_f: curve A shows the idealised density distribution across the image of a sharp edge, and curve B shows the practical density distribution on film (both enlarged in the x-direction)

more than twelve-fold, suggesting that film unsharpness and graininess have to be treated as two different parameters.

Granularity is a function of the film emulsion characteristics and the film processing whereas film unsharpness is primarily a function of the radiation energy absorbed in the film emulsion.

Table 3.4   Film unsharpness values for gamma-ray sources

Source	Unsharpness value (mm)	
Iridium–192	0.17	
Cobalt–60	0.35	
Ytterbium–169	0.07–0.13	} depending on the filtration
Thulium–170	0.10–0.20	or specimen thickness

An empirical relationship has been found between unsharpness, U, and resolving power, RP:

$$U(RP) = 2.9 \qquad\qquad (3.19)$$

where RP is in line-pairs per mm and U is in mm. Thus a conventional radiographic film used with $100\,kV$ X-rays may have a resolving power as high as 60 line-pairs per mm.

The films so far described are called 'direct-type' or 'non-screen type' radiographic films. They are exposed directly to X- or gamma-rays, or with metal intensifying screens (see Section 3.6.2). There is a second type of film, known as 'screen-type', which is used with fluorescent (salt) intensifying screens, which has quite different characteristics. Screen-type film is used for almost all medical diagnostic radiography, but rarely in industrial work. Basically, used with fluorescent screens, screen-type film has much poorer resolution (about 5–10 line-pairs per mm) and lower contrast, but a much higher speed. It reaches its maximum contrast at a density of about 1.3. Some manufacturers also sell 'wide latitude' films which have a lower value of gradient G and so can cover a larger specimen thickness within a given density range.

Film is processed in the appropriate proprietary solutions either by hand, or with automatic equipment. Some automatic processing equipment uses high

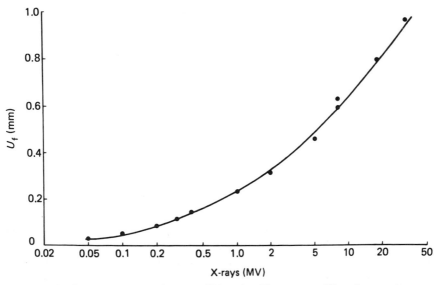

Fig. 3.26   Experimental curve of U_f against X-ray energy (filtered sources)

temperature solutions and can achieve a 'dry-to-dry' time of 5 minutes or less, with a large throughput. To obtain satisfactory results, it is advisable to use the solutions recommended by the film manufacturers, and there is no range of 'special' developers as there is in ordinary photography. A little can be gained by developing beyond the standard time, say 10 minutes instead of 5 minutes, but the real gains are marginal. There are some special applications which use non-radiographic and single-emulsion films, in order to obtain very-high-resolution images. The films used are the 'Line' and 'Process' types (see also Section 3.7.3).

Very recently, a 'dry X-ray film' has been introduced commercially. Processing is said to take 10 seconds and a resolution of 8 line-pairs per mm is claimed. The films are processed in a small bench-top roller-heater. They are normally used with rare-earth intensifying screens and are said to require exposure factors from $\times\frac{1}{2}$–$\times2\frac{1}{2}$ compared with Kodak M or Dupont NDT55 films, the factor depending on the radiation energy and the type of intensifying screen used. A maximum film gradient of 3.7 is claimed and a resolution of 8 lp/mm using a rare-earth salt screen.

Film radiographs must be placed on an illuminated screen to examine the image, and good viewing conditions in terms of the luminance level related to film density, low ambient viewing-room illumination, and masking to prevent glare, are absolutely essential. The ability of the human eye to discern both small detail and low contrast depends on the luminance being at least $30\,\mathrm{cd\,m^{-2}}$. For a film of density 3.0, an incident luminance of at least $30\,000\,\mathrm{cd\,m^{-2}}$ is therefore required if the eye is to see all the detail in the image. The setting up of good film viewing conditions is of vital importance to the success of film radiography: the eye must be able to see the detail which is present in the radiographic image.

Recently, attempts have been made to use closed-circuit television (CCTV) cameras for film viewing. Possible advantages are that a sensitive camera can handle very low brightness (i.e. high film densities); variable image magnification is easily possible; image processing can be used to provide image enhancement; quantitative image analysis is possible. Disadvantages are that the image is presented on a line raster; the dynamic range of a CCTV camera is limited; there may be noise added to the image.

3.6.2   Intensifying screens

Even with double-coated X-ray films, only a small fraction of the transmitted X-ray energy reaching the film is absorbed in the film emulsions, but this can be increased by employing intensifying screens.

In industrial radiography, these screens are usually thin lead foil, and a screen is pressed as closely as possible on each emulsion, so that there is a front and a back screen. The X-rays absorbed in the metal screen release electrons which pass into the film emulsion to add to the action of the directly absorbed radiation. The screens are very thin (0.02–0.2 mm) and of dense material, so that the electrons cannot spread sideways and the image unsharpness with a pair of screens is only very slightly poorer than with no screens. The intensifying action of the screens allows the exposure-time to be reduced by a factor between 2 and 6, depending on the X-ray energy. The use of metal screens has a secondary effect in preferentially absorbing the lower-energy

Compton scatter and so reducing the build-up factor, and thick screens are sometimes used for this purpose.

Radiographic film can be purchased ready-packed in envelopes with built-in lead intensifying screens. In some cases, the film and screens are vacuum-packed to ensure good emulsion/screen contact. Very thin (0.028 mm) lead screens are used.

For very-high-energy radiation, above 2 MeV, it has been found that materials other than lead produce the best results (see Section 3.10.3).

3.6.3   Fluorescent screens

There are materials which upon absorbing X-rays fluoresce in the visible and ultraviolet (UV) parts of the spectrum. A thin layer of a fluorescent powder in a binder can be used as an intensifying screen with a film which is sensitive to UV light (i.e. 'screen-type' films). Suitable materials for these screens are calcium tungstate and gadolinium oxysulphide. The image quality obtained with fluorescent screens is, however, very poor and is not considered adequate for most industrial applications.

Special screens, known as 'fluorometallic screens', consisting of a very thin coating of calcium tungstate on a lead backing are also available. The lead produces some electrons and the fluorescent layer produces UV light, so the photographic action is a combination of the two effects. To produce good radiographs, these screens must be used with special films which have some sensitivity, but not too much, to UV light. If the UV sensitivity is too high, the film/screen speed is high, but the image quality is poor. With the correct film, an intensifying factor between 3 and 10, compared with lead screens, is possible, together with good image quality. However, there are practical complications due to reciprocity law failure; furthermore, the screens are fragile and easily damaged.

If the fluorescence is in the visible part of the spectrum, the image on the screen can be seen directly by eye—the fluoroscopic technique—or 'seen' with a closed-circuit television camera—'television-fluoroscopy'—or the fluorescent screen can be built into an X-ray image intensifier tube (see Section 3.9). Suitable screen materials for fluoroscopy are zinc sulphide, cadmium sulphide, and caesium iodide.

3.6.4   Ionisation

Consider a beam of X-rays applied to two electrodes held a short distance apart (Fig. 3.27), with a potential difference, p.d., across them and an ionisable gas between them. When X-rays are absorbed, orbital electrons from the atoms of the gas are released and travel to the positive electrode, while the ionised atoms, having a positive charge, go to the negative electrode. Thus a current flows between the electrodes.

If the electrical potential is produced electrostatically, the X-rays cause a progressive loss of potential proportional to the energy absorbed.

If the p.d. across the electrodes is continuously maintained, the characteristics of the ionisation chamber vary with the applied voltage. The ionisation current can be measured as a direct current or as a series of pulses. In the ionisation region, each absorbed X-ray quantum produces a burst of electrons

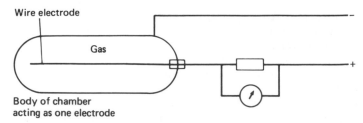

Fig. 3.27   Ionisation gauge for radiation intensity measurement

and, if the voltage is high enough to sweep these on to the anode before recombination occurs, the pulse size is independent of voltage: the ionisation current is proportional to the number of quanta absorbed in the chamber.

If the p.d. across the electrodes is then increased, the proportional region of the pulse/voltage characteristic is reached. Each electron produces more electrons by collision, so that both the pulse height and the number of pulses per second increase, and the size of the output pulse is now proportional to the input quantum energy.

At a still higher voltage, the proportionality between pulse height is lost and a region is reached, known as the Geiger region, in which the number of pulses per second is independent of voltage. The pulses are very large and the *number* of pulses is proportional to the incident radiation flux. Detectors used in this region are known as Geiger–Müller (GM) counters: they need very little amplification, and usually contain a self-quenching gas to prevent continuous discharge.

Ionisation devices are used for dose and dose-rate measurement, but the chambers cannot be made very small, so that to produce a two-dimensional spatial image of an X-ray field their use is limited and physically smaller phosphor/crystal detectors are preferred. Gas ionisation detectors have recently been built with up to 1000 elements in one gas-tight housing, and have a very high efficiency of energy conversion.

3.6.5   Scintillation counters

Certain phosphors absorb X-rays and convert the energy into light photons. If a photomultiplier tube is used with these, the individual light pulses can be detected, amplified and counted. The absorption of one X-ray quantum may lead to many light photons. The detector material may be a substance such as NaI (Th) which can be produced in bulk or in sheet form, or crystals such as CdS and ZnS. Using a large mass of a semi-transparent detector, the efficiency can be much higher than that of a gas-filled GM counter, as well as being inherently faster in response-time. Again, a scintillation detector can be used to count pulses or to produce a steady current output.

3.6.6   Semiconductor detectors

Certain materials such as cadmium sulphide change in conductivity under X-radiation, and this photoconductivity effect can be used to detect or measure X-rays. Light-sensitive photodiodes can be used in conjunction with a fluorescent screen. A scintillation crystal/photodiode detector array can be built as a linear array with sensitive elements as small as 0.1 mm wide, and such arrays are claimed to be capable of an image resolution of 4 line-pairs/ mm. Semiconductor junction devices with reverse bias behave much like a solid–state ionisation chamber; p–n and p–i–n junctions also show photoconductive effects.

As these are all small–element devices, they can be built into a matrix to detect the spatial X-ray image and there are two main ways of doing this. The first is to have a linear array of detectors which is mechanically scanned across the X-ray image, with its output fed into a television circuit on a slow-scan, line-by-line basis. The second method is to have a static, two-dimensional matrix of detectors. Both types are being developed to have digital outputs and can be used in television–fluoroscopic systems (see Section 3.9).

There are a number of other radiation detection devices.

It is possible to construct a panel consisting of a polycrystalline layer of a photoconductive material backed with a layer of electroluminescent material, and by applying a voltage across the double layer, make the panel act as a fluorescent screen with an enhanced output. The layers have to be thin, so that X-ray absorption is low, and the best amplification is obtained with high X-ray doses rather than low doses, which is the opposite of what is required in practice. Such electroluminescent panels have therefore not been much used in radiography.

Microchannel plates, MCPs, have been built which consist of a mosaic of very small cells of lead–glass tubing between two parallel plates, across which there is a high voltage (about 1 kV). The incident X-rays produce electrons in a front conversion screen which then produce electron avalanches in each cell of the microchannel plate by secondary emission, which is then converted to an image by an electron-to-light conversion screen (Fig. 3.28). Each channel is in

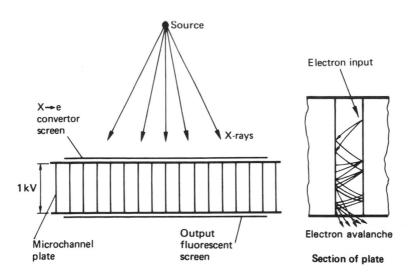

Fig. 3.28 Microchannel plate

the range 10–100 μm diameter and 2–3 mm long, and MCPs have been built with diameters up to 150 mm. The total gain can be of the order of 10^4.

If the surface of an insulator, such as a sheet of plastic, is given a uniform electrostatic charge of several hundred volts, and the sheet is placed in an X-ray field, the charge is destroyed locally according to the intensity of the X-rays, thus producing a spatial electrostatic X-ray 'image'. This image can be made visible by spraying with fine powder carrying the opposite electrical charge. This principle has been the basis of a number of processes of 'radiography' which have found limited use, two examples of these being xeroradiography and ionography.

Xeroradiography employs an amorphous selenium layer on a metal backing plate, and the image is formed with a sprayed-on very fine white powder. The plates can be cleaned, recharged and re-used. *Ionography* uses a thin plastic sheet on the inner surface of one electrode of a plate-shaped ionisation chamber. The electrons produced in the gas in the chamber discharge the electrostatic charge on the surface of the plastic sheet; the sheet is then extracted and the electrostatic image revealed by powdering (Fig. 3.29). There are several variations on these basic ideas, but none has yet found widespread use in industrial radiology.

Another possibility is the laser-stimulated-luminescence phosphor plate. A layer of europium-activated barium fluorohalide stores the absorbed X-ray energy in a quasi-stable state and the stimulation by light from a He–Ne laser causes the emission of luminescent radiation corresponding in intensity to the absorbed energy. This light is collected, intensified and its intensity is digitised. The laser scanning spot can be very small and this controls the image unsharpness. The light response to X-ray dose is linear over a very large range of intensities. After use, the phosphor plate is flooded with light and is then ready for re-use.

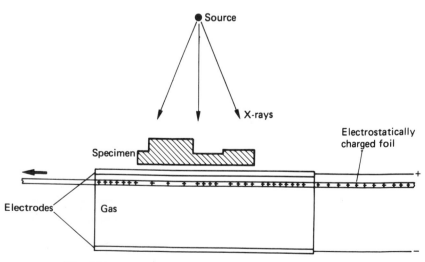

Fig. 3.29 Ionographic system for image on charged plastic foil

3.7 RADIOGRAPHIC TECHNIQUES

3.7.1 Basic principles—film radiography

The majority of specimens to which radiography is applied can be broadly divided into:

(1) approximately uniform thickness specimens, such as butt-welds, which are examined to detect internal flaws,
(2) non-uniform thickness specimens, such as small castings, also examined to detect internal flaws,
(3) assemblies, examined to check correctness of assembly, presence of specific components, spacing between components, etc.

In all these groups there may be a need for both high- and low-sensitivity techniques, but in groups 2 and 3 there is more emphasis on covering a specimen having a range of thickness—'latitude techniques'.

There are many parameters involved in detailing a radiographic technique, most of which have already been defined, and it is the purpose of this section to indicate how these are chosen and how they interrelate. Not all radiographs need to have the highest possible flaw sensitivity, but it is seldom that sensitivity is not an important factor.

Radiography is different from most NDT methods in that one equipment can be used for a very wide range of applications, with suitable choice of voltage, current, etc.

Thickness penetration

The first requirement is to have radiation which is capable of penetrating the specimen thickness, within a practicable, economic exposure-time. Table 3.5 gives a guide to the radiation required in terms of the performance of typical commercial equipment, film, etc.; together with standardised exposure charts

Table 3.5 Maximum thicknesses (mm)

Radiation	Steel (high-sensitivity technique)	Steel (low-sensitivity technique)	Aluminium (high-sensitivity technique)	Propellant (plastic) $\rho = 1.7$ g cm^{-3}
X-rays (kV)				
50	1	2	20	—
75	3	5	50	—
100	10	25	100	—
150	15	50	130	—
200	25	75	160	—
300	40	90	200	—
400	75	110	300	—
1000	125	160	—	500
2000	200	250	—	750
8000	300	350	—	1500
30000	325	400	—	2000
Gamma-rays				
Ir-192	60	100	—	—
Co-60	125	110	—	1000
Yb-169	8	12	30	—

(Fig. 3.30), which are usually supplied with individual X-ray equipment, a value of radiation can be determined. It should be emphasised that exposure curves of this type are drawn for a convenient standard source-to-film distance, and a particular film-type: this is not necessarily the correct source–to–film distance for good radiography. A conversion table for materials other than steel is also necessary (see Table 3.6).

At this stage, only an approximate value of the required X-ray voltage is needed. The second item of basic data needed is the effective focal-spot size of the X-ray equipment.

Using the approximate, appropriate voltage value, determined as above, the film unsharpness resulting from this X-radiation is read off Fig. 3.26.

Two decisions should now be made.

Firstly, if the requirement is for high-sensitivity radiography—e.g. butt-weld inspection, detection of fine cracks, or detection of small detail—it will be assumed that the radiographs will be taken on fine-grain, or very-fine-grain film, to a film density of 2.0. If fine cracks are not likely to occur and only grosser defects are likely, medium-speed film can be used, with shorter exposure-times, also exposed to obtain a film density of 2.0. Many applications standards, in fact, specify the type of film to be used.

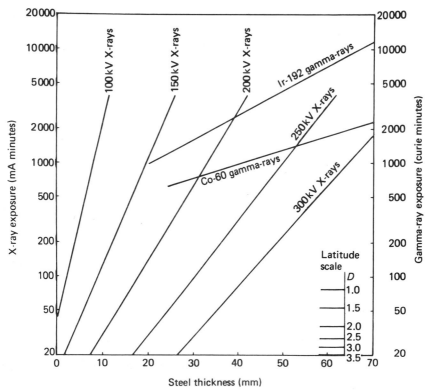

Fig. 3.30 Exposure curves for different X-ray energies, for steel plates, with D.4 film, thin lead screens, density 2.0, and a source-to-film distance of 1 m. Inset—a density scale to determine thickness latitude data

Table 3.6   Approximate equivalent thickness factors

Radiation	Steel	Aluminium	Magnesium	Copper	Lead
X-rays (kV)					
50	1.0	17.5	29.0	0.75	0.19
100	1.0	13.0	20.0	0.68	0.12
150	1.0	8.0	18.0	0.75	0.09
200	1.0	6.3	14.0	0.70	0.075
300	1.0	5.2	—	0.65	0.06
400	1.0	4.5	—	0.60	—
Gamma-rays					
Ir-192	1.0	3.0	—	0.90	0.25
Co-60	1.0	—	—	0.90	0.65

Note   Conversion values should be read horizontally, not vertically.

Secondly, if a high-sensitivity technique is required, the geometric unsharpness will be made equal to the film unsharpness,

$$U_g = U_f \qquad (3.20)$$

If a low-sensitivity technique is adequate, the geometric unsharpness can be allowed to be equal to twice the film unsharpness,

$$U_g = 2U_f \qquad (3.21)$$

The reasons for these last assumptions are that unsharpnesses are not directly additive:

$$U_T \neq U_g + U_f \qquad (3.22)$$

(U_T being the total effective unsharpness due to the two causes). The nett effect of the combination of two unsharpnesses is complex, depending on the precise shape of the unsharpness curve: one theoretical solution (Halmshaw, 1955) is

$$U_T = U_2 + U_1^2/3U_2 \qquad (3.23)$$

and $$U_T = (U_1^2 + U_2^2)^{1/2} \qquad (3.24)$$

is often taken as simpler and sufficiently accurate for most practical radiography (Klasens, 1946).

Having determined the appropriate value of geometric unsharpness and knowing the specimen thickness and the effective source diameter, the appropriate minimum source-to-film distance (sfd) is calculated from equation 3.6. The parameters of the appropriate technique are now known.

As a specific example, suppose it is required to radiograph a 30 mm steel weld: the available X-ray equipment will operate up to 300 kV and 10 mA, and has a 3 mm diameter effective focal-spot size. This application requires a high-sensitivity, fine-grain film technique. For 30 mm steel, from Table 3.5 and Fig. 3.30, the required X-ray voltage will be in the range 250–300 kV. From Fig. 3.26, this means that $U_f = 0.12$ mm. So, from $U_g = U_f$ and equation 3.6,

$$0.12 = sb/a = 3 \times 30/a$$

$$a = 750 \text{ mm}$$

Thus the minimum source-to-film distance which should be used is $(750 + 30) = 780$ mm.

The curves of Fig. 3.30 are drawn for a fine-grain film, density 2.0, which is what is required here, but for 1 m sfd; so the exposure required can be corrected to 780 mm by equation 3.15, to give

at 300 kV and 10 mA $t = 1.76$ minutes (106 s)

at 250 kV and 10 mA $t = 5.80$ minutes

Many codes of good radiographic practice (such as BS:2600 and BS:2910) give *minimum* source-to-film distances calculated in this manner, and some further examples for both X- and gamma-rays are listed in Table 3.7.

The sfds given in Table 3.7 are the *minimum* values that need to be used to obtain good image sharpnesses, but it is emphasised that still higher quality radiographs are possible by taking $U_g = \frac{1}{2}U_f$. These values clearly show that the use of a standard source-to-film distance of 100 cm is unjustified. In many applications, a much shorter sfd is completely satisfactory.

Investigation of the effect of sfd on crack sensitivity has shown that the falloff with reducing sfd is at first gradual and that a reduction in the values of Table 3.7 by 20% can often be acceptable: this would result in an exposure reduction of 36%.

For irregular-thickness specimens, this procedure needs to be slightly extended. The thickness of the thickest part of the specimen is taken and the appropriate values of voltage, U_f, U_g, and sfd are determined in the same way. From the *smallest* thickness in the specimen, using the appropriate line on the exposure chart and a correction factor from Table 3.8, the exposure-time is determined for a film density of 3.5 (this value being usually taken as the highest film density which can be viewed satisfactorily).

Table 3.7   Minimum source-to-film distances, based on $U_g = U_f$

Radiation	Source diameter (mm)	Specimen thickness (cm)	Minimum sfd (cm)
X-rays (kV)			
100	1.0	0.5	10.5
150	2.0	2.0	60.0
150	2.0	10.0	300.0*
250	4.0	2.5	92.0
400	5.0	7.0	240.0*
1000	7.0	10.0	294.0*
8000	2.0	25.0	125.0
Gamma-rays			
Ir-192	0.5	1.0	5.0
Ir-192	2.0	5.0	82.0
Co-60	2.0	10.0	67.0
Co-60	4.0	10.0	123.0
Yb-169	0.5	0.5	4.5†
Yb-169	0.3	0.2	1.4†

*Only with these values are the exposure-times likely to be embarrassingly long, and it is only with these techniques that it may be necessary to devise a compromise between a slightly poorer sensitivity and a shorter exposure-time.

†These values show the application of this gamma-ray source to very small diameter pipe-welding (see also Section 3.10).

The exposure-time for the thickest part of the specimen is then read off the same line of the chart, and the ratio of the two exposures can be used with Table 3.8 to determine the film density under the thickest part of the specimen. If this density is less than 1.0, Table 3.8 shows that there will be a large loss of contrast in this part of the image and the radiograph may not be acceptable.

A convenient method of determining thickness latitudes in relation to X-ray energies, is to use the exposure-factors of Table 3.8 on a scale which can be moved across exposure charts such as Fig. 3.30. A scale of this type, to be used as a moveable cursor, is shown in Fig. 3.30.

There are several techniques for producing improved results on irregular-section specimens, as follows.

(1) Expose together two films of different speeds, the faster film to cover the thickest sections.

(2) Use a higher than normal X-ray voltage.

(3) Use a filter close to the X-ray tube to remove some of the softer components in the X-ray spectrum and so 'harden' the beam. A combination of filters close to the X-ray tube and thicker than normal front metal intensifying screens is frequently used for difficult specimens. For filters close to the tube, typical thicknesses are:

with 200 kV X-rays   0.25–0.50 mm lead

with 400 kV X-rays   0.60–1.00 mm lead

Filters are very rarely used with high-energy X-rays or with gamma-rays.

(4) Use a film with a smaller contrast-gradient: for extreme cases, a single-emulsion film can be used.

Table 3.8 Contrast loss and exposure-factors for typical fine-grain radiographic film

Film density	Contrast as % of value at density of 2.0	Relative exposure required taking $E_{2.0}$ as 1.00
0.5	0.22	0.22
0.8	0.41	0.36
1.0	0.50	0.47
1.3	0.65	0.62
1.6	0.78	0.78
2.0	1.00	1.00
2.5	1.24	1.29
3.0	1.43	1.55
3.5	1.61	1.74

Note These values apply to one typical radiographic film but will vary only slightly with other films.

3.7.2   Scattered radiation

With X-rays in the energy range 150–400 keV, where Compton scatter is predominant, large amounts of non–image-forming scattered radiation can reach the film and spoil the image. As can be seen in the basic image contrast equation (see equation 3.14), if I_S/I_D is large, the contrast is greatly reduced. There are various techniques to alleviate the problem:

(1)   to use a front metal intensifying screen, slightly thicker than normal,
(2)   to use pre-shaped masks or a diaphragm to limit the area being radiographed,
(3)   to use edge-masking with lead sheet, lead paste, or fine lead shot—these techniques can be extremely tedious and should be treated as a last resort,
(4)   to use heavily-filtered, higher-energy radiation.

Very-high-energy radiations (above 1 MeV) and gamma-rays rarely require any masking techniques.

A technique which virtually eliminates scattered radiation, but which requires special equipment, is the projective magnification method (Fig. 3.31). The specimen is placed close to the X-ray source and the film is at a distance behind. Since the image is projected several times natural-size, it would be blurred unless a very small X-ray source is used, so this is a technique limited to microfocus X-ray tubes. The advantages are:

(1)   the effect of scattered radiation is almost eliminated; if $M > 3$,

$$(1 + I_S/I_D) \rightarrow 1$$

(2)   since the images of flaws in the specimen are enlarged, the effects of film unsharpness and film graininess on the image are much reduced and better flaw sensitivities are possible.

The disadvantages of projective magnification methods are:

(1)   the need for special microfocus X-ray tubes,
(2)   the low X-ray output of these tubes, requiring longer exposures,

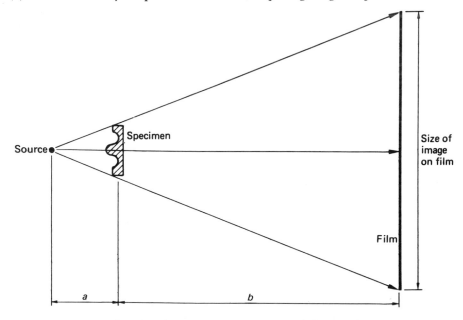

Fig. 3.31   Principle of projective magnification:

$$\text{magnification, } M = \frac{a+b}{a} = 1 + \frac{b}{a} = 1 + \frac{Ug}{S}$$

(3) the greater the magnification that is used, the smaller is the area of specimen covered.

Various anti-scatter grids, both static and moving grids, such as are used in medical radiography, are rarely if ever used in industrial techniques.

Scattered radiation can produce many forms of spurious image, and if unexplained dark patches or bands appear on the radiograph, it is a good *a priori* assumption that these are caused by scatter.

3.7.3 Special high-resolution techniques

For small specimens in which very small detail needs to be discernible, there are two possible techniques.

The first technique uses Line or Process film instead of conventional X-ray film. These are single-emulsion, high-contrast films designed for lithographic work and have a very fine grain. They can be processed in conventional X-ray developers for 3 minutes instead of 5 minutes and then have a moderate contrast ($G = 2.5$ at density 2.0) and a reasonable density range. These films have a very high silver/gelatine ratio, so they have a smaller unsharpness with X-rays, and when used with a technique having a low value of U_g, they can be enlarged optically up to about $\times 30$ before the graininess and unsharpness begin to have a serious effect on image detail. Large numbers of small assemblies can therefore be radiographed simultaneously on one film and the images examined optically under a magnifier or by projection.

The second technique is to use a microfocus X-ray source and projectively magnify the image of the specimen directly on to the film as described in Section 3.7.2. Then the geometric unsharpness, related to the specimen detail, is the actual geometric unsharpness divided by the magnification, M. The effective geometric unsharpness, U_g, is given by

$$U_g = s(M-1)/M$$

For example, a source of 5 μm can be used with a magnification of $\times 21$ to produce a magnified image whose unsharpness is 0.1 mm (equal to the film unsharpness with 200 kV X-rays). Higher magnification with this focus size would produce a less sharp image.

Projective magnification methods on film therefore require very small focal-spot sizes if they are to be used with very high magnifications, and also have the disadvantage that only one or two specimens are imaged at each exposure; the large images require large-area films. However, on thick specimens, there is no problem with geometric unsharpness due to the specimen thickness, and there is minimal scattered radiation.

The special film technique described above can be used with any X-ray equipment and X-ray energy, whereas the microfocus technique requires a special X-ray tube.

3.7.4 Gamma-radiography

The same procedure for calculating the appropriate source-to-film distance as described for X-rays, is used with gamma-rays, but because it is not possible to vary the energy from a source, the initial choice of appropriate source is made from the specimen thickness and the ranges shown in Table 3.9.

Table 3.9 Steel thickness ranges for which gamma-ray sources are suitable

Source	High-sensitivity technique (mm)	Low-sensitivity technique (mm)
Ir-192	18–80	6–100
Cs-137	30–100	20–120
Co-60	50–150	30–200
Yb-169	2–12	1–15
Tm-170	2–12	1–15

The lower end of the thickness range is governed by the loss of image quality (flaw sensitivity), and the upper end by the long exposure-time required. The metal intensifying screens used with Ir-192 and Cs-137 are the usual thin lead screens, but with Co-60 gamma-rays, thicker copper or steel screens give better results. With Yb-169 gamma-rays on thicknesses less than 5 mm, the lead screens should be omitted: this lengthens the exposure-time, but produces better quality images.

3.7.5   Stereoscopic methods

A pair of radiographs can be taken with the X-ray source shifted in a plane parallel to the film between the two exposures. These two films can be examined in a stereoscopic viewer, and if there are suitable markers on the two surfaces of the specimen, a three-dimensional image will be seen, enabling the depth of an internal flaw to be estimated.

A more common technique to determine the depth of a flaw is to make measurements on the two films, or on a double exposure on one film (Fig. 3.32). There are several variants in the calculation procedure according to whether or not the specimen thickness is accurately known. Distances A, B and C are measured on the film and are plotted on a graph (Fig. 3.32(b)), with $y = aA/h$: this gives points P, Q and R, and d is a measure of the distance from the specimen surface to the flaw. The shift distance h is kept as small as will give a reasonable distance for C and A.

3.7.6   Determination of the through-thickness dimension, h

An extension of the stereometric method, to determine the through-thickness dimension of a crack, is possible. Two images are taken, with a small source shift at right angles to the length of the crack. These can be either two separate films or two images on one film.

The *width* of each of the crack images is then measured (not the separation). If the widths of the two images are w_1 and w_2, the height of the crack, h, is given by

$$h = \tfrac{1}{2}(w_1 + w_2)\cot\theta/2$$

where θ is the angle between the two X-ray beams. θ is assumed to be small, and the source-to-film distance large, compared with h. It has been found that the best method of measuring w is to project the radiograph on to a screen at about $\times20$–$\times30$ magnification, with a high-quality projector. The accuracy of measurement of h has been found to be slightly better than ±1 mm over a wide range of specimen thicknesses and radiations.

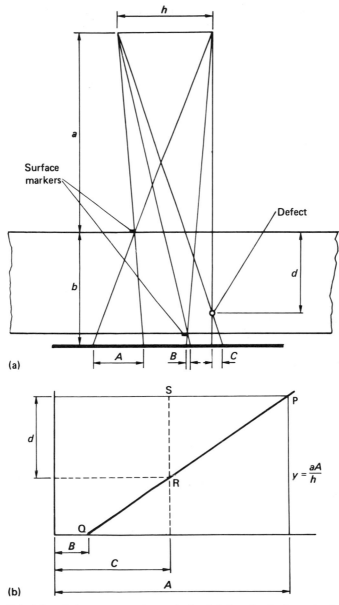

Fig. 3.32 Principle of the stereometric method of defect depth determination from two exposures

The through-thickness dimension can also be determined from a density measurement, if a microdensitometer scan across the image of the defect can be obtained. Microdensitometer traces are usually linear with density, so the density difference between the defect image and the local background, ΔD, is easily obtained.

Either a thickness–density curve is required, which can be obtained from a densitometer scan over a small step-wedge laid on the specimen, or the film characteristic and exposure curves can be used. From the film characteristic a linear relationship can be assumed:

$$\Delta D = G_D \, \Delta(\log E)$$

The obtained value of $\Delta(\log E)$ is transferred to the appropriate exposure curve, to determine $\Delta t = h$. The use of a step-wedge calibration eliminates the need to assume a value of G_D, but it is necessary for the film density outside the step-wedge to be the same as the background film density close to the image of the cavity being measured.

3.8   RADIOGRAPHIC SENSITIVITY

A distinction should be made between radiographic quality and radiographic sensitivity, although for many applications, such as flaw detection, the two are virtually synonymous: the ability to show a smaller flaw (i.e. an improvement in sensitivity) is usually regarded as an improvement in quality. In the radiography of assemblies, however, detail sensitivity may be relatively unimportant. In industrial radiography, 'sensitivity' nearly always means the ability to show detail, and the term is not used for the speed of the system.

As a consequence of the large numbers of parameters in a radiographic technique—choice of film type, voltage, source-to-film distance, film density, etc.—a wide range of sensitivity values is possible and ever since the earliest days of radiography, devices called 'penetrameters', now known as 'image quality indicators, IQIs', have been used to measure sensitivity.

The earliest penetrameters were simple step-wedges, made of the same material as the specimen, and placed on the source side surface of the specimen, so that their image appeared on the radiograph. The thinnest step of the wedge, the image of which could be discerned on the radiograph, was taken as a measure of the thickness sensitivity. This is the sensitivity described by equation 3.14 and is a measure of image contrast. It is a useful value, but it gives no information on image definition. Later designs of penetrameter involved drilling holes or patterns of holes in the steps, but few of these designs have survived and today there are basically three designs of IQI in general use, with a fourth, newer design which has at present only a limited use.

The ability to detect a flaw in a casting or a weld depends on the nature of the flaw, its shape, and orientation to the radiation beam, as well as on the technique parameters, so that IQI sensitivity is not directly a measure of flaw sensitivity, although a relationship between the two can be established (see Section 3.8.6). Different designs of IQIs do not give the same numerical value of IQI sensitivity on the same radiograph, so it is essential to quote the type of IQI being used in specifying required or attained values.

3.8.1   Wire Type IQI

This design (Fig. 3.33(a)), which is standardised in the UK, in Germany, the Netherlands, Scandinavia, and recently in the USA; also by the International Standards Organisation and the International Institute of Welding, consists of

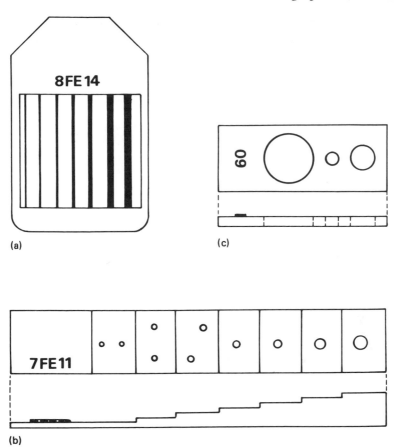

Fig. 3.33   Image quality indicators: (a) wire type, (b) step/hole type, (c) plaque type

straight wires, 30 mm long, of the same material as the specimen, with wire diameters selected from the geometric series in Table 3.10. A typical wire IQI will contain seven of these wires.

The wires are held in a low-X-ray-absorbent sheet material with appropriate identifying lead letters. In the German (DIN) design, the wire diameters are the same as in Table 3.10, but the wire *numbers* start at the thickest wire ($d = 3.20$ mm being wire number 1).

The wire IQI is put on the surface of the specimen facing the radiation source and the thinnest discernible wire on the radiograph is taken as a measure of the wire IQI sensitivity.

The IQI sensitivity can be expressed as an actual wire diameter, or more commonly as a percentage value:

$$\text{wire IQI sensitivity} = \frac{\text{diameter of smallest discernible wire}}{\text{thickness of specimen, under wire}} \times 100\%$$

Expressed as a percentage number, a smaller numerical value means that a smaller diameter wire has been detected, and, by assumption, a smaller flaw

Table 3.10   Wire IQI: wire diameters (tolerance ±5%)

Number	Diameter* (mm)
1	0.032
2	0.040
3	0.050
4	0.063
5	0.080
6	0.100
7	0.125
8	0.160
9	0.200
10	0.250
11	0.320
12	0.400
13	0.500
14	0.630
15	0.800
16	1.000
17	1.250
18	1.600
19	2.000
20	2.500
21	3.200

*These diameters correspond to Standard Wire Gauge diameters.

would be detected: i.e. the sensitivity is said to be 'better' or 'higher' although the numerical value is lower.

3.8.2   Step/hole type IQI

A step-wedge (Fig. 3.33(b)) of the same material as the specimen is used, and in each step there are one or two drilled holes of the same diameter as the step thickness. The step-thicknesses follow the same series as the wire diameters of Table 3.10, but the thinnest step (step number 1) is 0.125 mm. The step/hole IQI is used in the same manner as the wire IQI and IQI (hole) sensitivity is taken as the diameter of the smallest discernible hole. This design of IQI is used in France, where it originated, and is also specified in BSI and ISO standards and by AFNOR.

3.8.3   Plaque type IQI

These are widely used in the USA and are still called penetrameters (Fig. 3.33(c)). There are several designs, but basically, all consist of a uniform thickness plate of the same material as the specimen, which is placed on the specimen surface. The thickness of this plate, T, should be 2% of the specimen thickness, and each has three drilled holes, of diameters T, $2T$ and $4T$. Occasionally, 1% and 4% plaques are used, and also $2T$, $3T$ and $4T$ holes. Lead identification numbers are put on each penetrameter.

On most specifications of the plaque type penetrameter, although the holes are considered as T, $2T$ and $4T$, there is an overriding limit that the minimum hole diameters are 0.01″, 0.02″ and 0.024″, respectively (ASTM), and in another design (ASME) a minimum of $\frac{1}{16}''$. On thin specimens, less than 1" (25 mm) thick, there can therefore be considerable confusion as to what is actually to be discerned on the image when a $2T$ hole is specified.

The penetrameter sensitivity with the plaque type can be determined in several ways.

(1) The discernibility of the plaque (not the holes) gives a measure of contrast sensitivity.

(2) The 'level of sensitivity' obtained or required can be quoted in the form $(2-2T)$, $(1-2T)$, $(2-1T)$, etc., where, for example, $(2-2T)$ level signifies that the $2T$ hole in a 2% thickness plaque shall be discerned.

(3) 'Equivalent' sensitivity values may be used (Table 3.11). These values originate from an arbitrary assumption that $(2-2T)$ shall be called 2% equivalent and that the equivalent values for other hole sizes are proportional to $(TR)^{1/2}$, where R is the hole radius. This can lead to considerable confusion, as a $1T$ hole in a 2% plaque is called 1.4% sensitivity in the USA, but it would be called 2.0% (hole) sensitivity in Europe.

Table 3.11 ASTM levels of sensitivity

Level	Equivalent sensitivity (%)
1-1T	0.7
1-2T	1.0
2-1T	1.4
2-2T	2.0
2-4T	2.8
4-2T	4.0

3.8.4 IQI sensitivity data

None of the above patterns of IQI is entirely satisfactory. There are problems of deciding on what is meant by 'just discernible' on an image, particularly with a wire, where the image tends to become discontinuous before actually disappearing. There are also problems in making special IQIs for materials other than steel and aluminium, and in accuracy of construction. Nevertheless, in many radiographic specifications the quality of the radiograph required is specified by an acceptable IQI value. As an example, for butt-welds in critical structures, one standard specifies that the values given in Table 3.12 shall be obtained. These values represent reasonably good quality radiography, but they can be improved by careful choice of techniques.

With regard to Table 3.12, the following points should be noted.

(1) It is clear that a single acceptable value of sensitivity cannot be proposed: for example, 1.6% (wire) would represent a very good radiograph on 6 mm steel, but a very poor one on 100 mm steel.

(2) The column of values in Table 3.12 represents a smooth curve, but when using an actual wire IQI, only specific wire thicknesses are available: on 40 mm, for example, the actual IQI sensitivity readings obtained would be either 1.25% or 1.0%, rather than the 1.1% listed.

(3) The change in IQI value, with a change in radiographic technique, is not very great: these designs of IQI are not particularly sensitive to definition

Table 3.12 IQI values which should be obtained with a satisfactory radiographic technique

Steel specimen thickness (mm)	IQI sensitivity (%)	
	Wire IQI	Step/hole type IQI
3	2.4	5.1
6	1.6	3.6
12	1.4	3.0
25	1.2	2.5
40	1.1	2.1
50	1.0	1.8
75	0.9	1.6
100	0.8	1.4
150	0.7	1.3

parameters. As an indication of the degree to which these values can be exceeded by high quality techniques, wire IQI values on 146 mm steel of 0.43 and 0.54% were claimed in a recent paper, using linac and cobalt-60 gamma-rays respectively.

3.8.5   Duplex-wire type IQI

A completely different pattern of IQI has been developed by CERL/UK and is now also included in BS:3971. Each element (Fig. 3.34) consists of a pair of straight wires spaced one diameter apart, and made of a very dense material such as tungsten or gold. Judgement is based on the finest pair of wires which can be seen on the radiograph *as a pair* of wires and not merged. The diameter of the wires in the first merged pair is then a measure of the total unsharpness of the image. The wire pairs range from 0.05 mm diameter to 1.6 mm diameter, and there is also a model of this IQI containing pairs of bars, based on the same principle, for use with higher-energy radiation.

The smallest wire pair corresponds to a total unsharpness value of 0.10 mm, which means that present models of this IQI are not suitable for high-definition techniques on thin specimens. Also, this type of IQI is difficult and fairly expensive to fabricate, so it has not yet come into widespread use. In spite of these limitations it has some important technical advantages over the better-known designs of IQI.

3.8.6   Detail sensitivity

It must be emphasised that IQI sensitivity is a measure of the quality of a radiographic technique and not a direct measure of flaw sensitivity. Neverthe-less, relationships can be developed between IQI and flaw sensitivities.

If one considers a very small cavity of width W and height b in a specimen of total thickness t, the excess radiation penetrating the cavity is

$$I_0 W \{\exp[-\mu(t-b)] - \exp(-\mu t)\}$$

which is approximately equal to

$$I_0 \mu (Wb) \exp(-\mu t)$$

Detail of round wires

Fig. 3.34   Duplex-wire type image quality indicator

then, for a small wire of radius r, the change in quantity of direct radiation reaching the film per unit length of wire is

$$\pi r^2 \mu I_D \tag{3.25}$$

In the presence of unsharpness, U, the image of a cavity or a wire is broadened and, if $U > W$ (or $2r$), there is also a reduction in contrast (Fig. 3.35); C (the actual contrast) is less than C_0 (the contrast when $U < 2r$).

The width of the image of the wire on the film, if $U_T > 2r$, is

$$(2r + U_T) \tag{3.26}$$

and combining this with equations 3.14 and 3.25 gives

$$\frac{r^2}{2r + U_T} = \frac{2.3\Delta D(1 + I_S/I_D)}{G_D \mu \pi} \tag{3.27}$$

From this equation, r can be evaluated and the wire IQI sensitivity is $[(2r/t) \times 100]\%$.

The density distribution in the image of a wire is not uniform (Fig. 3.35), and so it has been suggested that the normal 'minimum discernible density difference, ΔD', will not be the same as for a simple density-step, and a 'form-factor' should be introduced, writing ΔDF instead of ΔD. For small wires, $F = 0.6$; for large wires, $F = 1$. Calculated and experimental values of wire IQI sensitivity have been found to be in good agreement, suggesting that film graininess, which is not part of the equation, is not an important factor in wire discernibility.

A similar calculation can be used to develop a formula for crack sensitivity:

$$dW = \frac{2.3\Delta DF}{G_D \mu}(d\sin\theta + W\cos\theta + U_T)(1 + I_S/I_D) \tag{3.28}$$

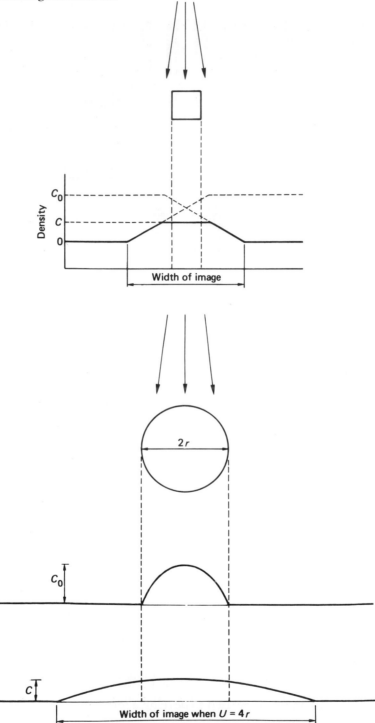

Fig. 3.35 Effect of unsharpness on the image of a small defect

where W is the opening width of the crack, d is the height, and θ is the angle between the crack and the radiation beam (Fig. 3.36).

In Fig. 3.36(a), the 'crack' is drawn with a uniform opening width W, but as a natural crack will taper towards its propagating tips, it is more realistic to take the opening width as $W/2$ as in Fig. 3.36(b).

In so far as it is possible to measure the dimensions W and d of any natural crack, experimental measurements of crack sensitivity agree well with calculated values. Equations 3.27 and 3.28 can also be combined to produce a relationship between wire IQI and crack sensitivity. For cracks aligned with the X-ray beam ($\theta = 0$)

$$d = \frac{3U_T r^2}{(2r + U_T)W} \quad \text{when } U \gg W \tag{3.29}$$

This relationship is not therefore a direct one, but depends on the acting total unsharpness: thus two radiographic techniques giving the same wire IQI sensitivity do not necessarily give the same crack sensitivity. This is easily confirmed by varying the geometric unsharpness: the wire IQI sensitivity changes only very slightly, whereas the crack sensitivity varies much more.

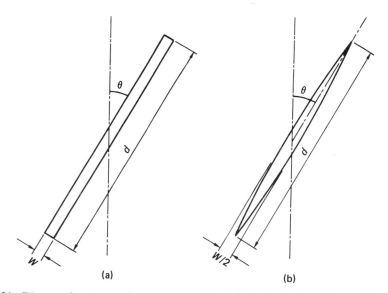

Fig. 3.36 Diagram showing crack parameters d, W and θ: (a) idealised parallel-sided slit and (b) tapered crack

3.8.7   Flaw sensitivity

There is therefore no simple relationship between IQI sensitivity and flaw sensitivity and it is certainly not true that if a 1% wire IQI sensitivity is obtained, all flaws occupying more than 1% of the specimen thickness are detectable: it is reasonable to suggest, however, that some relationship exists.

There are five commonly-occurring types of flaw.

(1)   *Small air-filled pores or gas holes*   These occur in welds and castings and are usually approximately spherical in shape. They therefore roughly match the drilled hole in a step–hole IQI, with relative volumes in the ratio 2:3, so that if a step/hole IQI sensitivity of 2.0% is obtained, the gas-hole sensitivity will be 3%.

(2)   *Piping*   This is a linear discontinuity of roughly circular cross-section, so it is in effect a negative-thickness wire: wire IQI sensitivity will therefore directly measure piping detectability.

(3)   *Lack of penetration at the root of a weld*   This is analogous to a zero-angle artificial crack (see equation 3.28).

(4)   *Large-area cavities*   If the cavity is much larger than the acting unsharpness, the sensitivity is largely a question of contrast, and corresponds to the simple thickness sensitivity.

(5)   *Cracks*   If these are very narrow, with a large height-to-width ratio, there is no IQI with corresponding detail, and in any case the detectability depends as much on angle θ (Fig. 3.36) as on W and d. If the acting unsharpness is known, there is a relationship between crack sensitivity and wire IQI sensitivity (see equation 3.29).

The position of the flaw through the thickness of the specimen has some effect on flaw sensitivity. Geometric unsharpness is smaller for flaws close to the film, but with a good technique based on $U_g = U_f$, this factor is not of great importance as U_T will be approximately equal to $1.3U_f$. Due to scattered radiation, a flaw on the film surface of the specimen is slightly enhanced in contrast, but this effect is minimised by the effect of the front intensifying screen.

3.8.8   Interpretation

A most important part of the radiographic inspection is the 'reading' of the radiograph. The importance of appropriate film viewing conditions has already been emphasised and the desirable screen luminances, masking conditions, etc., specified. Automated image recognition systems, using computer programs for pattern recognition are still only in the early stages of development, so that film interpretation depends on the skill and experience of the inspector or radiologist.

Four stages of interpretation are necessary:

(1)   to verify that the film corresponds to the specimen and that the technique specified has been used—the latter point is usually verified by the IQI sensitivity,

(2)   to recognise any artefacts or spurious marks,

(3)   to identify the nature, location and dimensions of any flaws in the specimen,

(4)   to report the results.

Successful interpretation requires a knowledge of the design of the specimen and the manufacturing processes involved, training and experience, as well as some intuition. For example, a radiologist should not attempt to interpret weld radiographs without knowing the weld preparation shape, the number of runs, and the X-ray beam direction.

There are collections of 'Reference Radiographs' for welds and castings (ASTM and IIW) which can be used for comparison purposes. These are valuable for training and for standardisation of nomenclature, but when considering the range of specimen thicknesses, materials, radiographic techniques, and the variety of applications, it is rare that a flaw found on a radiograph can be exactly matched to a radiograph in a reference collection. In these reference radiograph collections there are usually several levels of severity of each flaw: for example, very severe, severe, moderate, slight, very slight porosity. For spatially-distributed flaws, such as porosity, microshrinkage and microporosity, the illustration of these by films rather than words is quite convenient, but it must be remembered that a severe example on one application might be considered an unimportant flaw for a less critical application, and any acceptance/rejection sentencing must be based on a knowledge of the specific application of the specimen.

Interpretation can best be taught by tutorials with an experienced radiologist and a collection of radiographs, but a few fundamental points can be listed.

(1)   In radiography, an air-filled cavity represents a reduction in specimen

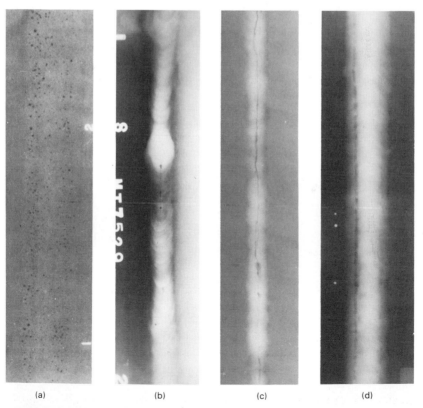

(a)   (b)   (c)   (d)

Fig. 3.37 Radiographs of butt-welds in steel plate, taken with X-rays: (a) 20 mm plate (dressed-weld) with weld containing distributed small porosity, (b) 8 mm plate (as-welded) containing a short length of root penetration and associated gas holes, (c) 6 mm plate (as-welded) containing a longitudinal crack in weld metal, and (d) 15 mm plate (as-welded) containing a linear slag inclusion on the side wall

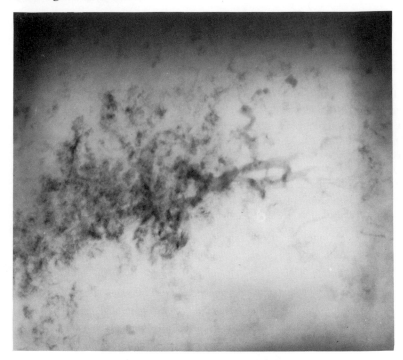

Fig. 3.38 Radiograph of cast steel, 50 mm thick, showing filamentary shrinkage, cavities, and shrinkage piping

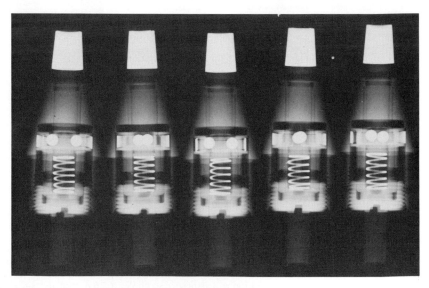

Fig. 3.39 Radiograph of a row of small ordnance shell fuses, taken to show internal assembly details

thickness in that area, so that more radiation is transmitted to the film, so producing a darker (higher-density) image. A lower-density image means extra thickness or high absorption.

(2)   Except with projective magnification techniques, the film images are within 10% of natural size.

(3)   The image of a flaw is the projection of a three-dimensional cavity on to the plane of the film. There is one important exception to this: the images of very narrow flaws can be spread out to several times their true width by unsharpness effects (see equation 3.26).

There is no room in this volume for an atlas of radiographs, and, in any case, paper reproductions of film radiographs are usually unsatisfactory. However, a few examples of radiographs are shown in Figs 3.37, 3.38 and 3.39.

3.9   FLUOROSCOPIC METHODS

If a fluorescent screen, which converts X-rays to light, is put behind a specimen, an image of the specimen can be seen on this screen. The image is usually very faint, so that fine detail cannot be discerned and flaw sensitivities are usually poor. The method, using a mirror-viewing system for safety, is still used for low-density items, such as letter packets and plastic components, but rarely for metal specimens. A suitably sensitive closed-circuit television (CCTV) camera can be focussed on the fluorescent screen, and the amplification circuits associated with the camera used to produce a bright image on a television monitor screen. This type of equipment will be referred to as Group 1 under the general heading of 'television-fluoroscopic systems' (Fig. 3.40). Since the image is presented on a television monitor, it can be remote from the X-ray equipment and all radiation hazards are eliminated. For most applications, a CCTV camera tube with an extra intensification stage, such as a Sitcon or an Isocon, must be used to obtain the necessary light sensitivity.

X-ray image intensifiers were first marketed about 1950, primarily for medical use. They consist (Fig. 3.41) of a fluorescent screen backed with a light-to-electron conversion screen at the output end, both screens being in a vacuum tube. The electron image from the double-layer input screen is accelerated and focussed by electron lenses on to the smaller output screen, and this electron acceleration plus image minification produces an image intensification factor of between 100 and 1000. The output screen can be viewed directly with a magnifier, but it is much more convenient to use a CCTV camera, and present the image on a monitor: this is the Group 2 type of television-fluoroscopic system (Fig. 3.42). A simpler CCTV camera, such as a vidicon type, has adequate sensitivity because of the light intensification in the X-ray image intensifier; solid-state cameras are also used. Modern X-ray image intensifier tubes are built with thick CsI front screens, specially made for the higher X-ray energies used in industry, and tubes are available up to 40 cm diameter.

These two groups of television-fluoroscopic systems have been available for about twenty years but have found only limited application in industrial work because the image is formed on a polycrystalline conversion screen, with an

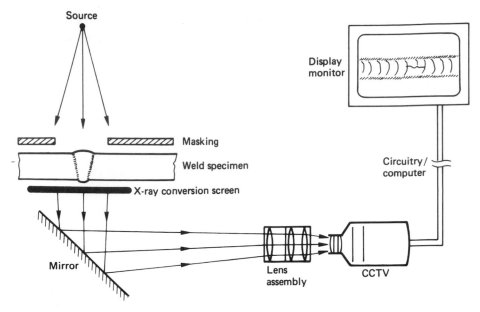

Fig. 3.40   Television-fluoroscopic system (Group 1)

unsharpness in the range 0.6–1.0 mm, and this is not sharp enough for many applications. They have been used successfully for the inspection of light-alloy castings, some composite structures, and for some medium-speed ciné-radiography applications. In recent years, however, there have been three major developments which can greatly increase the performance of these systems and make them suitable for a much wider range of applications.

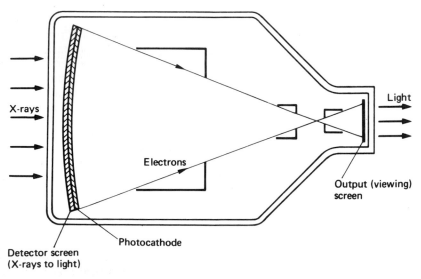

Fig. 3.41   X-ray image intensifier tube

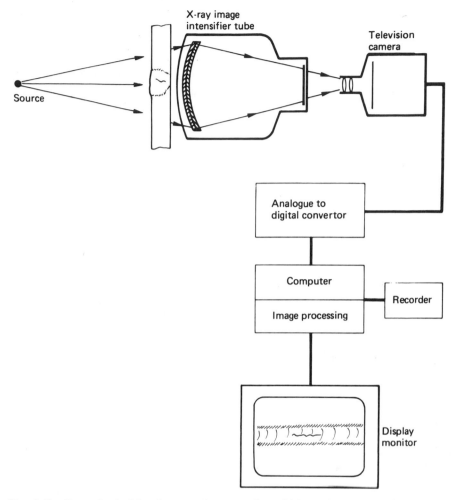

Fig. 3.42 Group 2 television-fluoroscopic system for weld inspection, using an X-ray image intensifier tube

Digital imaging

A television image can be converted into digital data and this data held in a digital framestore. Then, with a random-access memory and a computer, digital image processing and enhancement is easily possible. Small computers now have the capability of rapid processing of this quantity of data and the term 'real-time radiography, RTR', has been coined for these techniques. 'Radioscopy' is a more logical name and is increasingly being used.

Pixels

Any image can be regarded as a series of very small picture elements—pixels—and, as a television image is formed on a line raster, it is common practice to take an image as 512 lines, with 512 picture elements per line ($512 = 2^9$): each element is assigned a brightness (grey scale) value on an 8-bit scale, i.e. 256 levels of grey. Obviously, the size of a pixel is a function of the overall image size: on a 300 mm wide image, each pixel would be 0.6 mm wide; on a 100 mm image, only 0.2 mm wide. 512×512 pixels therefore can give a reasonably detailed image; 256×256 pixels would lead to more rapid processing; 1024×1024 pixels at present would take too long to process and require a much more expensive computer. There is little to be gained in working with a very large pixel matrix, unless the performance of the other components in the system—the camera, the line raster, and the bandwidth—are also upgraded.

Any television–fluoroscopic system should be analysed in terms of its frequency response—modulation transfer function (MTF) values—throughout all its transfer stages, to produce an overall performance curve.

Some television cameras will produce a digital output directly; with others an analogue-to-digital converter is necessary.

Another recent development is the 'solid-state' television camera. These use a two-dimensional array of light-sensitive elements behind a lens, each element of the array having a digital output on a progressive scanning circuit. There is therefore no electron scanning beam in these cameras. The sensitive elements can be either charge coupled or complementary metal-oxide semi-conductor devices and arrays of 1024×1024 and larger are now available. At present, these cameras have light sensitivities similar to vidicons, and so can be used with Group 2 systems, but not with Group 1 systems. They are very robust and small, and have much better performances, in terms of absence of lag and a larger dynamic range, than conventional cameras.

Projective magnification

The third development in television–fluoroscopic systems is the use of projective magnification (p.m.). This technique has already been briefly explained in terms of film radiography of very small specimens.

If the primary conversion screen of a fluoroscopic system is placed at a distance behind the specimen, the image is enlarged, and if the focal spot of the X-ray tube is small enough, the image can still be sharp, geometrically. Detail on the specimen will also be enlarged, so that the effect of the screen unsharpness and its graininess on the image of the detail is reduced. The effective unsharpness, referred to the specimen detail, is

(total unsharpness)/(projective magnification)

Microfocus X-ray sets, with focal spots as small as 5 μm are available, which permit large values of M, and for some applications a minifocus tube ($s = 0.3$ mm) and a small amount of p.m. can be shown to be useful.

One disadvantage of using p.m. is that as M is increased, the area of specimen covered at one position is reduced by $1/M$, which inevitably slows down the speed of inspection.

3.9.1   Quantum fluctuation limitations

X-rays are emitted, absorbed and scattered as X-ray quanta, and Planck's quantum theory states that the processes must be regarded as having a random statistical nature. An X-ray beam can be described as containing A quanta $mm^{-2} s^{-1}$, but this does not mean that in each mm^2 of the beam, there are exactly A quanta s^{-1}. If a pixel in the image is formed from, *on average*, the absorption of N quanta, there is a natural fluctuation in the value of N, which is $N^{1/2}$. The signal-to-noise ratio is therefore

$$N/N^{1/2} = N^{1/2}$$

It can be shown that this same condition applies at all stages of the imaging process—primary screen, electron conversion, television signal, etc.—and that the quantum noise at the stage at which the number of quanta is the smallest (which is almost always at the absorption of the X-ray quanta in the primary conversion screen) will dominate the noise level in the final image. It can be shown that most television-fluoroscopic systems, even those with only a moderate amplification factor, are, in practice, quantum-noise limited (Sturm & Morgan, 1949). The performance of any imaging system—television camera, film, human eye, image intensifier, etc.—can be described in terms of its quantum-limited performance (Rose, 1948).

With a television system, the image is renewed every 1/25 s, so that for small-area pixels, the number of X-ray quanta utilised to form each pixel can be quite small: the image will therefore be noisy, and this image noise may obscure image detail.

There are three ways of alleviating this limitation:

(1) to devise a primary conversion screen which will absorb and convert a greater proportion of the incident X-ray quanta (special screen materials and thicker screens are possible: mosaic screens of scintillating glass have been fabricated for this purpose),

(2) to use a higher output of X-rays or a higher X-ray energy, to produce a higher X-ray intensity on the screen,

(3) to integrate the signal over several seconds, instead of using only the number of X-ray quanta absorbed in the time of one television frame (this method can only be applied to a static image, but is relatively easy to do with a digital system of image storage).

There are several available computer procedures, and the ability to integrate and average over several television frames is one of the most important advantages of using digital techniques. It is possible to average over 8, 16, 32 or more television frames in a very short time, so the delay in 'seeing' an image need be only a few seconds. The signal-to-noise ratio can be shown to be proportional to $F^{1/2}$, where F is the number of frames used.

3.9.2 Digital image processing

Image quality can be described in terms of contrast, definition, and noise. With frame integration, the noise problem, so far as quantum noise is concerned, can be virtually eliminated. Contrast modification, both overall and local, can be very easily achieved with quite simple computer programs. True image sharpening is a little more difficult to achieve, but there are several suitable computer programs, through Laplace transforms, edge enhancement, deconvolution, and spatial filtering: at present, most of these tend also to enhance the noise.

A four-stage image enhancement program is frequently used:

(1)   field equalisation,
(2)   frame integration and averaging,
(3)   image crispening,
(4)   contrast enhancement,

and improved computer programs in this field are developing very rapidly. Filtering techniques can be applied in either the spatial or frequency domain.

Other image modification procedures are also possible, such as subtraction techniques (subtracting a standard image, to leave only abnormalities on the displayed image), pseudo-colours, edge-outlining, spatially-displaced subtraction, and pattern recognition. Image compression techniques such as those used in medical diagnostic radiography, have not yet been applied to industrial radiographs, but may be needed for economic digital recording. All these processes can be hand-controlled through a keyboard, or hard-wired for speed of operation, or automated.

3.9.3   Other equipment

Apart from Groups 1 and 2 equipment, it is possible to use a linear array of small detectors as the primary X-ray absorber/converter (Fig. 3.43). This array is moved across the specimen, or in some applications the specimen moves across a fixed array on a travelling belt. The latter method is widely used in airport luggage-search equipment: the X-ray beam need be only a narrow slit, 3 mm wide, so that scattered radiation is greatly reduced, with an improvement in image quality.

A typical array may consist of 600 X-ray-sensitive elements, each about 3×1 mm, consisting of a photodiode covered with a layer of fluorescent material (see Section 2.5). Each element of the array produces a digital output, which is time-sequenced by a shift register, and taken into a framestore, as one unit of pixel data.

Digital radiographic imaging is an increasingly important field, particularly in medical diagnostic radiography, and is developing rapidly. There are a number of possibilities.

A flying spot of X-rays can be generated by a rotating radial slit crossing a fixed slit and if this is incident on a long detector crystal, the travelling spot can be time-controlled to produce a raster of pixels; by moving the specimen on a belt across the X-ray beam, a matrix of pixel data is generated. So far this method has been applied only to airport luggage-search equipment, but industrial equipment has been proposed with a 4096×4096 matrix of pixels on a 20×20 cm imaging area, which would give a spatial resolution of 0.05 mm. The data from the scanning beam can be programmed to produce tomographic images from the same equipment. The use of such a large pixel matrix will require very high data-capacity image processing facilities.

Other possibilities for X-ray television-fluoroscopic systems include light amplifier tubes, which can be used to increase the light intensity on to the CCTV camera tube. Microchannel plates have been used for the same purpose. 1100- or even 2000-line raster television systems can be used.

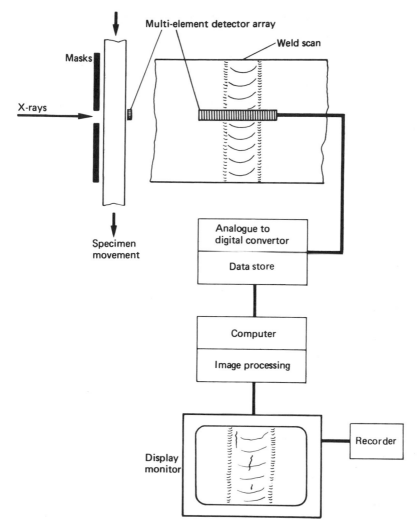

Fig. 3.43   Group 3 television–fluoroscopic system for weld inspection, using a linear array of detectors

3.9.4   Performance

Signal digitisation and the use of framestores, computers, and digital image processing has given a very large boost to X-ray television-fluoroscopic methods. With appropriate techniques, IQI sensitivity values equivalent to those attainable in film radiography can be obtained, but it must be remembered that the formation of the image is quite different. On a television system the image is presented on a display monitor, on a line raster, and may be considerably enlarged to minimise the effect of the lines on the image detail: this is therefore quite different from the slightly grainy image on a radiographic film.

It was shown in Section 3.8.6 that conventional IQIs are much more sensitive to contrast than to definition parameters and this is reflected in the good IQI values attainable with television-fluoroscopic equipment.

To obtain equivalent values of crack sensitivity (equation 3.28), it is necessary to reduce the overall unsharpness of the system to the unsharpness values attainable with the same radiation energy on film, and this can only be done with projective magnification (p.m.). Taking the total unsharpness of a television-fluoroscopic system as 0.5 mm (equivalent to a limiting resolution of 5 line-pairs per mm), this leads to the p.m. and source-size values listed in Table 3.13.

Table 3.13   Projective magnification and source sizes needed to make $U_g = 0.6U_T$, to match film radiography

Radiation	Specimen thickness (mm)	Projective magnification	Focal-spot or source size (μm)	Source-to-detector distance (mm)
100 kV X-rays	5	10.0	3	650
200 kV X-rays	20	5.0	15	640
400 kV X-rays	50	2.5	80	640
Ir-192 gamma-rays	50	2.0	150	520
Co-60 gamma-rays	100	1.0	—	520

Many potential applications of X-ray television-fluoroscopic systems do not necessarily require a high crack or IQI sensitivity.

The concept of having a radiographic image in digital data form will probably lead to automated inspection equipment with pattern-recognition programmes in the not-too-distant future.

3.10   APPLICATIONS

Industrial radiography has a very wide range of applications and it is not possible to detail these in a small space. A few applications will be described to illustrate special points of technique.

3.10.1   Pipe welding

Pipelines are fabricated by butt-welding lengths of pipe, using 360° circumferential butt-welds. Some joints may be made in a machine shop, or a fabrication yard, and some are made on site, or, for undersea pipelines, on a pipelaying 'lay barge'. For some site work, the rate of welding, and therefore the rate of inspection required, can be very rapid—as short as 3 minutes per weld. The welds are inspected either by radiography or by ultrasonic testing: in either case, a full 100% inspection is usually required. Pipes vary from 48" diameter down to 6" diameter for pipelines, and down to 1" diameter for chemical plant, but the commonest sizes are in the 6"–16" range.

There are four basic methods of radiographing a circumferential butt-weld (see Fig. 3.44).

With method (a), the radiation source is at the centre of the pipe, and if it is a panoramic emitter, a film can be wrapped all round the weld and the

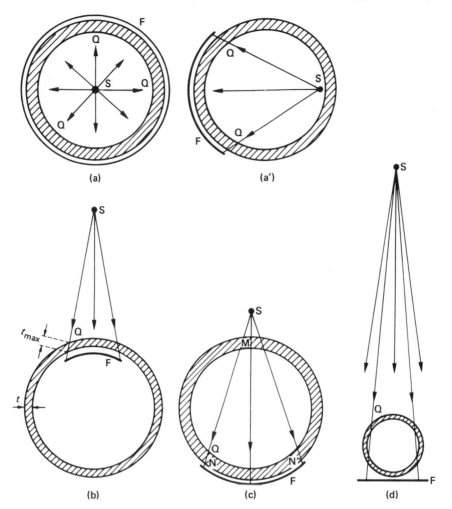

Fig. 3.44   Radiography of circumferential butt-welds in pipes: (a) source inside (central), film outside, single wall method, (a′) source inside (off-centre), film outside, single wall method, (b) source outside, film inside, single wall method, (c) source and film outside, double wall, single image method, and (d) source and film outside, double wall, double image method. F and S represent the film and source respectively, and Q shows the position of the image quality indicator

examination made in one exposure. This is therefore a very economic method, but depends on being able to position a source of suitable physical size to give the required value of U_g and to have the necessary energy to penetrate the weld thickness. Figure 3.44(a′) shows a variant on this technique, if the dimensions of the available source are too large to use satisfactorily on the pipe axis. In this case, two or three exposures are necessary. Remote-controlled 'crawlers' are made to carry radiation sources (both X- and gamma-ray) long distances along the bore of a pipe and locate the source accurately on the weld axis. Crawlers to go into pipes as small as 6″ ID (inner diameter) have been built, and some crawlers carry a battery-pack to power the X-ray set.

In method (b), it is necessary to place the film cassettes inside the pipe. At each position, the thickness of weld penetrated at the edge of the film is greater than at the centre, so the film density decreases towards the ends of the film and there is a limit to the length of weld which can be satisfactorily examined at each position. A common criterion is that

$$t_{max} \not> t_{min} + 10\%$$

where t is the penetrated thickness. From this, curves can be derived showing that the number of films to cover a weld can vary between about 5 and 15 according to the thickness/diameter ratio. This is not therefore a rapid inspection technique, even when the pipe can be rotated under a fixed source.

Method (c) does not require access to the inside of the pipe, but the radiation must penetrate two walls to obtain an image of the weld. The weld at M is so far from the film that its image will be extremely unsharp and interpretation is limited to the weld length between N and N'. Since there is extra absorption in the radiation beam, compared with the single wall methods (a), (a') and (b), the flaw sensitivity related to the weld thickness will be considerably poorer than with the single wall methods.

Method (d) is only applicable to small diameter pipes. If a small diameter source is used at a large source-to-film distance, the image definition of the weld image on both sides of the pipe may be acceptable.

All these methods may be used with gamma-rays, as well as with X-rays, if a source to match the thickness ranges given in Table 3.9 is available. In method (c), the gamma-ray source is often put very close to the pipe surface at M.

Television–fluoroscopic methods have been proposed for pipe-welding inspection, using methods (a), (b) and (c). In large pipes, the detector assembly or the radiation source can be inside the pipe; in smaller pipes, it can be rotated in steps around the pipe.

With methods (a) and (c), a technique problem arises in that the correct location for an IQI is point Q, and this point may be completely inaccessible. A control exposure should be made on a separate piece of pipe of the same size, to check the adequacy of the technique parameters.

3.10.2   Very-low-energy X-rays

If X-rays below 40 kV are used, some special factors need to be considered.

The X-ray tube should have a special window, usually thin beryllium, to reduce the inherent filtration in the X-ray beam. The conventional X-ray cassette with a thin metal or plastic front should be replaced by one having a thin, black polythene front (0.1 mm). Below 20 kV, the air absorption between the X-ray tube and the specimen begins to be significant and the air should be replaced with a less absorbing gas or a vacuum. The most practical techniques are to use either a helium-filled balloon between the X-ray tube and the specimen, or a helium-filled drum with foil end-faces.

As specimen thicknesses are very small, geometric unsharpness is rarely an important factor, but as the discernible image detail size is about 2% of the specimen thickness, film graininess is important, and very-fine-grain films or single-emulsion films should be used. A front metal intensifying screen would act only as a rather heavy filter and should rarely be necessary, but a back screen can be retained to minimise back scatter.

The change in absorption with voltage is very rapid at energies below 40 kV, and a high-output X-ray source which allows a further slight reduction in voltage can produce considerable improvements in detail sensitivity. Thus the value of μ for aluminium changes by almost a factor of three between 25 and 20 kV X-rays. At still lower voltages, the value of μ for polythene changes from 10 to 18 between 10 and 7 kV X-rays. Choice of correct voltage can therefore be quite critical.

A method of enhancing crack detection sensitivity, which has been used in low-voltage radiography of composites, is to immerse the specimen in an X-ray-opaque penetrant, which soaks into any cracks. Excess penetrant is wiped off before radiography. Halogenated hydrocarbons such as trichloro-trifluoroethane have the highest opacity, but inorganic salts such as zinc iodide, lead nitrate, barium sulphate may also be used. A similar procedure, using carbon tetrachloride on aluminium alloy castings, has been reported.

An interesting application of very-low-voltage X-rays is to radiograph a film radiograph. If a radiograph is taken of a specimen with a very large thickness range and an exposure is made for the thickest areas, the thinner parts will be grossly over-exposed and impossible to 'read'. The resulting film can be used as a specimen and radiographed with 7–20 kV X-rays (depending on the type of film) using a short source-to-film distance, to produce a readable image of all the thicknesses in the specimen, while retaining acceptable image contrast and definition.

3.10.3 Megavoltage X-radiography

The equation for thickness sensitivity, equation 3.14, contains the ratio (build-up factor)/μ which is sometimes called the 'radiation contrast factor', and if this ratio is reduced, the image contrast increases. With X-ray energies above about 2 MeV and steel thicknesses greater than about 70 mm, this ratio will be found to be reducing, because of the faster reduction in the build-up factor than in the value of μ. There is therefore the prospect of obtaining a higher contrast image by using a higher X-ray energy.

The output from most megavoltage X-ray equipment, such as linacs, is very large, so that large source-to-film distances can be used with short exposure-times and fine-grain films. Consequently, the geometric unsharpness is nearly always small and total unsharpness is controlled by the film unsharpness. Source-to-film distances are chosen in terms of the required field size, rather than the geometric unsharpness. With energies above 4 MeV, there is strong evidence that metal intensifying screens of materials other than lead will give better-quality radiographs. With 5–8 MV X-rays, thick copper intensifying screens, 1.0–1.5 mm thick, have been shown to give the best radiographs for many applications, but the problem has not been thoroughly investigated and the conclusions are probably dependent on the scatter conditions both in the specimen and around it. With 8–30 MV X-rays, a front screen of 1.0–1.5 mm of tungsten or tantalum is recommended, with no back intensifying screen, unless there is known to be considerable back scatter on to the cassette.

In megavoltage radiography, therefore, inherent film unsharpness is a major factor in determining image quality, and there might be a good technical case for the use of special films having a higher silver/gelatine ratio, and so a smaller value of U_f.

3.11   SPECIAL TECHNIQUES

Images can be produced with atomic particles such as neutrons, protons, and electrons. As these are partially absorbed during their transmission through matter, and an image can be obtained on a film, the techniques are commonly called 'radiography'.

3.11.1   Neutron radiography

Neutrons can have a range of energies from a fraction of an electron-volt to several MeV. 'Thermal neutrons', with energies between 0.01 and 0.30 eV, and 'cold neutrons', with energies between 0 and 0.01 eV, have some properties which make them especially interesting for radiography. Whereas the attenuation of X-rays in materials increases with atomic number in a reasonably orderly manner, the mass absorption coefficients of the elements for thermal and cold neutrons appear almost completely random. Thus hydrogen, lithium, boron, gadolinium, dysprosium, indium, and cadmium have very high absorption coefficients, whereas aluminium, iron, and lead have coefficients 500–2000 times smaller, the reason being that neutron absorption does not depend on the electronic structure of the atom, but on interactions with the atomic nucleus.

The most important case is hydrogen, and the high absorption coefficient implies that small amounts of hydrogenous material (e.g. water, oil, plastic, and explosive) can be detected by radiography with neutrons through considerable thicknesses of metal such as steel or aluminium.

Neutron images can be recorded on conventional photographic film by five techniques.

(1)   By direct exposures using metal foils on either side of the film during exposure: foils of thin gadolinium, cadmium, or indium are used.
(2)   By direct exposures using screens of a material such as Li-6 mixed with ZnS: these screens convert absorbed neutrons to light; they are very fast, but give grainy images.
(3)   By direct exposures using a scintillator glass screen: these are made of a glass incorporating Li-6, and also emit light; they are slower, but produce less grainy images than polycrystalline screens.
(4)   By a transfer technique: a foil of dysprosium or indium is exposed behind the specimen and develops a spatial radioactive image; this activated foil is then placed on a film in a cassette, and left to expose the film as the radioactivity decays.
(5)   By a 'track etch' technique: a sheet of cellulose nitrate is pressed against a converter screen containing B^{10}. The converter screen emits α-particles on absorption of neutrons, and these cause tracks in the cellulose sheet which are etched out by immersion in caustic soda solution. The image is sharp, but of low contrast. The etched image can be seen by eye, but is better shown by photographing the sheet with appropriate lighting. Track-etch sheets produce high-resolution images, but have a low speed.

The major problem in neutron radiography is the availability of sources of neutrons, and while there are several potential sources of intense neutron beams, nearly all produce fast neutrons which must be moderated to produce thermal or cold neutrons. This moderation process is very inefficient in terms of producing a small, intense source of thermal neutrons, which is the requirement for radiography, the losses being typically $1:10^6$. Possible neutron sources are:

(1)   atomic reactors,
(2)   'neutron tubes'—positive ion generators, employing the T(d, n) reaction in a vacuum tube,
(3)   high-energy X-ray machines using an (X, n) conversion process in a metal such as beryllium or uranium (i.e. absorbing the high-energy X-rays in the metal, when fast neutrons are released),
(4)   radioactive sources using the (α, n) or (γ, n) reactions,
(5)   radioisotopic neutron sources.

In practice, nearly all neutron radiography sources are atomic reactors, because these have the ability to produce an intense thermal or cold neutron beam. With sources of fast neutrons, these are thermalised by multiple collisions in a moderator consisting of wax, oil, water, or polythene (Fig. 3.45) and the thermal neutron beam is extracted by a collimator tube of either cylindrical or conical form: the length/diameter (L/D) ratio of this collimator is equivalent to the (source-to-film distance)/(source diameter) ratio in conventional radiography. A large atomic reactor can produce a thermal neutron

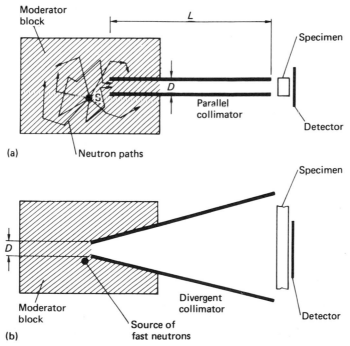

Fig. 3.45   Neutron radiography—thermalisation and collimation of beam for: (a) a parallel-sided collimator and (b) a divergent collimator

(n) intensity at the film plane of 10^8 n cm^{-2} s^{-1} with an L/D ratio of 200, whereas other sources are likely to be no better than 10^5 n cm^{-2} s^{-1} at an L/D ratio of 20. Several small reactors have been specially designed for neutron radiography.

As an alternative to atomic reactor sources, there is one neutron–emitting radioisotope, californium-252. This is a transuranic element, made artificially, and a 5 mg source has an output of 11.5×10^7 n s^{-1}. Again, these fast neutrons have to be thermalised and the thermal neutron output is likely to be about 1×10^4 n cm^{-2} s^{-1} at an L/D ratio of 20. Californium-252 has a half life of 2.65 years. In the USA, a Cf-252 source with a flux multiplier has been investigated. By placing the source near the centre of an array of enriched uranium fuel rods in a water moderator, multiplication of the neutron flux by a factor of 30 has been achieved.

The use of a high-power X-ray linac to produce neutrons by absorption of the X-rays in beryllium, uranium, or lead is attractive in that the main cost of the machine is as megavoltage X-radiographic equipment, and conversion to neutron generation need only be done when specific applications arise. The neutrons still need to be moderated and great care is necessary to shield the films from the intense X-ray beam. Using a 12 MeV linac, a uranium block and a divergent collimator, an output of 1.10^6 n cm^{-2}s^{-1} can be obtained at the film plane.

It has been estimated that 10^5 n cm^{-2} on a detector are needed to produce a reasonable image, and 10^9 n cm^{-2} are needed for a high-quality image, and with all sources except the large atomic reactors a compromise is necessary between L/D ratio, exposure-times, and the image sharpness; geometric unsharpness is likely to be the predominant cause of loss of image quality.

Image intensifier tubes with neutron-sensitive conversion screens have been built, and a few applications using low-intensity neutron beams have been reported. With image integration, this technique appears to have considerable potential.

With cold neutrons, the images obtained with metal foil screens by the direct method show an outlining effect on some image detail which enhances detectability: the effect is probably due to very-low-angle neutron scatter.

There are possibilities in enhancing sensitivity by loading porous materials with a gadolinium nitrate solution, or by mixing in a boron compound to increase neutron absorption. One successful application was to absorb gadolinium nitrate solution into any residual mould cores left in the cooling ducts of cast turbine blades, to enhance detection of the mould residues. Neutron radiography of thick materials has not found widespread application, possibly because in thick materials there can be a secondary generation of gamma-rays from the absorbed neutrons, leading to considerable scattered radiation and, consequently, poor flaw sensitivity.

Neutron radiography of atomic reactor fuel elements is a very important application, where good radiographs can be obtained in spite of the high gamma-ray background. This can be done underwater, where the detector and source are positioned by remote control in a container, and then the water is removed by compressed air for the duration of the exposure. Track–etch detectors are usually used.

3.11.2   Computerised tomography

Several new imaging processes have been developed from the basic concept of measuring the radiation intensity at a series of points across an X-ray image, digitising the data, and using a computer program to reconstruct an image.

The most important of these methods, originally developed by Hounsfield (1979) for medical applications, is computerised axial tomography (CAT or CT). Industrial applications are developing rapidly.

The specimen is scanned (Fig. 3.46(a)) by a narrow beam of X- or gamma-rays, which impinges on to a linear array of detectors. The X-ray source, the beam collimator, and the detector assembly are fixed to a common frame which rotates round the specimen in a series of small steps. At each step, the output of each detector is read, digitised and stored, so that each reading represents the absorption, or the transmitted intensity, along one narrow line through the specimen. Hounsfield developed equations and solutions to convert the readings into an X-ray image of a narrow axial slice through the specimen: the reconstruction algorithms were based on convolution and back projection techniques; iterative reconstruction algorithms are also possible. The image obtained is as if a slice has been cut through the specimen and this slice laid on a film for radiography: it is therefore an entirely different image to a conventional radiograph on which the images of objects at different depths are superimposed.

For industrial applications, it is often more convenient to rotate and translate the specimen, with a fixed radiation source and detector array (Fig. 3.46(b)): other configurations of source and detectors are possible. The time to collect

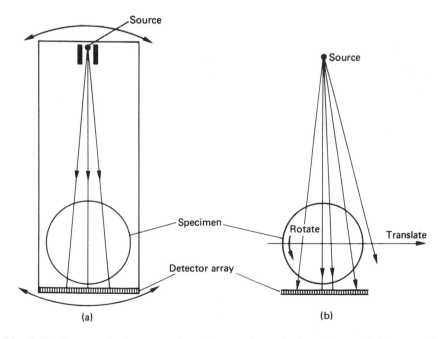

Fig. 3.46 Computerised tomography: (a) set-up for a fixed specimen and (b) set up for a specimen with rotational and translational movement, with fixed source and detectors

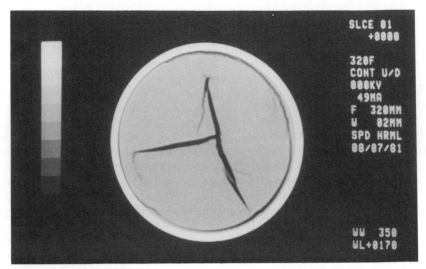

Fig. 3.47   Computerised tomography of explosive filling in metal casing, showing very severe cracking and interface defects

the X-ray data is usually only a few seconds, but data processing can take considerably longer unless a large computer or an array processor is used, or a small pixel matrix is acceptable.

Apart from the X-ray CAT image, very sensitive absorption data is produced as a byproduct of the computations: physical density measurements to an accuracy of 0.02% are possible.

CAT techniques have been applied to such varying objects as large solid–propellant rocket motors, in situ tree trunks, ordnance shells (Fig. 3.47), and multistrand cable, and both techniques and applications can be expected to develop rapidly. Back scatter tomographic techniques, with a radiation source and detector on the same side of the specimen, have been proposed. Computerised tomography with gamma-rays is also possible.

3.11.3   Compton scatter methods

An array of detectors can be used to measure the radiation transmitted through a specimen, and the readings obtained used in a computer to construct a conventional radiographic image. The method eliminates the use of film, film processing and film storage, and permits methods such as image subtraction and image enhancement to be used. It is sometimes called 'digital radiography', but has not yet found much use industrially.

If the off-axis Compton scatter is collected by a collimator, or a series of collimators and detectors (Fig. 3.48), these readings can also be used to obtain a computer-developed image of the specimen. Obviously, the amount of material in the radiation beam will affect the amount of scatter generated, and theory suggests that by measuring this scattered radiation it may be possible to detect flaws in solid specimens with a sensitivity as good as, or even better than that with conventional film radiography, and at a higher inspection speed. Rayleigh scattering depends very strongly on the X-ray energy and on the

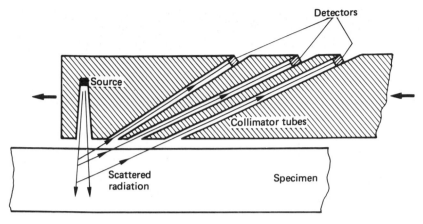

Fig. 3.48   Compton scatter system, with source and detectors on the same side of the specimen plate

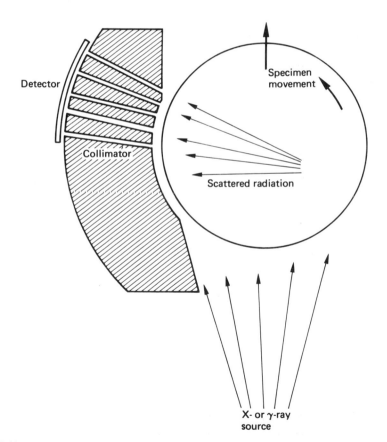

Fig. 3.49   Compton scatter system for the examination of large, filled ordnance shells

atomic number of the scattering material. Only at very low energies is the scatter at a large angle.

The theory involving multiple scattering processes, is exceedingly complex, and has to relate radiation source intensities, collimator widths, scanning speeds, radiation collection times, and scatter intensities to acceptable signal-to-noise ratios.

With Compton scatter the effects of multiple scattering can be eliminated by using an energy-sensitive detector.

Compton back scatter equipment (Fig. 3.49) has been built for the automatic inspection of large filled ordnance shell; it is claimed to have a resolution of 1 mm and a high inspection speed. It uses a high energy linac as radiation source. Similar equipment has been proposed for the inspection of underwater off-shore structures, using a cobalt-60 gamma-ray source: the equipment would require access to one side only of the specimen (Fig. 3.48), which is a great advantage in many site applications. Most of these back scatter systems have used small diameter holes to collimate the scattered radiation, with the consequent need for high-intensity sources and long exposures.

A different technique has been proposed in US Patent 422 9651:1980 for specimens of low-density material. This method proposes a slot camera (Fig. 3.50) as detector, with either film and fast intensifying screens, or an array of radiation detectors. In preliminary investigations on multilayer rocket chamber walls, gaps between layers, delaminations, and volumetric defects have been detected. Gaps between layers as small as 0.15 mm and larger gaps up to 70 mm below the surface have been imaged, using 200 kV X-rays as a primary source. Exposure-times have been 20–120 minutes, but it is believed that with a detector array, real-time quantitative inspection can be achieved.

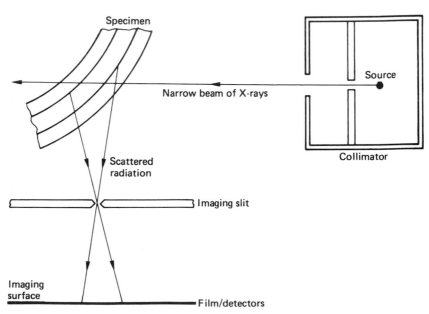

Fig. 3.50 Proposal of Compton scatter system for rocket-motor casing inspection

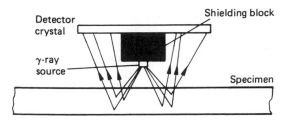

Fig. 3.51   Back scatter thickness gauge

The slot camera has the advantage that it images the complete wall thickness in one view, and unlike normal radiography requires access to one side only.

Back scattered radiation can also be used for thickness measurement (Fig. 3.51).

3.11.4   Moving-beam systems

A radiograph can be taken by scanning a narrow beam of X-rays along the specimen, by means of a narrow slit (Fig. 3.52). The method has the following advantages.

(1)   The beam forming the image is always at right angles to the specimen, so avoiding the image distortion present at the edges of a large film. This is particularly important on specimens such as large honeycomb structures, or on specimens where it is desired to make measurements of distances.
(2)   The X-ray scatter is minimised by the small beam width.
(3)   Protection of personnel is much easier because of the small area of specimen being irradiated.
(4)   The image is obtained on one sheet of film.

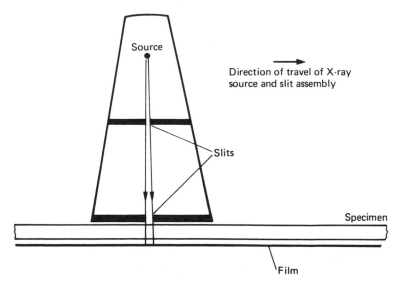

Fig. 3.52   Moving beam (dynamic) radiography, with travelling slit and fixed specimen and film

The disadvantages are that, except for very thin specimens, there is some blurring of image detail. This can be made smaller by using a narrower scanning slit, but this increases the total exposure-time as a slower scanning speed is then necessary. The image blurring in the direction of motion is related to the slit width rather than to the source diameter, but there are complex relationships on the discernibility of very small object detail, which are related to the source size as well as to the slit width and the specimen thickness.

Circular in-motion radiography on a cylindrical specimen is possible and can also be used with a linear detector array as a television-fluoroscopic slow-scan system. Typical in-motion film radiography systems operate in the speed-range 5–15 cm per minute.

3.11.5   Flash radiography

X-ray image intensifier tubes can be built with a primary conversion screen having a short decay-time, and if a ciné camera is optically coupled to the output screen, ciné-radiography frame-speeds up to about 300 frames per second can be used. The X-ray equipment will need to have a smoothed, continuous output, unless the ciné camera is synchronised to the X-ray pulse rate.

There is, however, a need in scientific applications for very-high-speed ciné-radiography and also for very-short-exposure radiography (exposure-times in the range 20 ns to 1 μs), to study such phenomena as explosive events. The usual problem is to get a very high X-ray output in each pulse.

There are several ways of producing short pulses of X-rays:

(1)   by discharging a large capacitor circuit through the X-ray tube,
(2)   by discharging a large stack of piezoelectric discs,
(3)   by using a pulse generating circuit,
(4)   by electronically 'chopping' a longer pulse.

In addition, some X-ray machines, such as linacs and betatrons, produce their X-ray output in pulses.

For single pulses the basic principle used is the discharge of a bank of capacitors or a storage circuit through a special form of X-ray tube. In these tubes a filament is not used and electrons are generated at a specially-shaped cathode by field emission. The instantaneous tube current can be 10^3–10^4 amperes, with pulse-times in the range 10^{-6}–10^{-8} s. With high voltage tubes outputs up to 300 roentgens per pulse at 1 m distance have been claimed. Special 'flash linacs' have been built with even higher outputs.

Field emission is a process whereby electrons are emitted when a very high electric field (50 MV cm^{-1}) is applied to a metal surface. The field is concentrated at the tip of a needle, or array of needles, which forms the cathode of the X-ray tube. The tube target is usually conical to reduce the source diameter.

The main applications of flash radiography have been in military ballistics and explosives research, so that much of the development is classified information. The attraction is that images can be obtained in spite of the smoke and flash which limit the use of flash photography. The X-ray equipment must of course be protected from the explosive event, so it is often necessary to take the radiographs through protective armour. As a consequence, most flash radiography uses very–high–speed salt intensifying screens and fast films: often a multiple sandwich of films and screens is employed and the films superimposed for viewing.

It is possible, by using several flash X–ray tubes, to record the image on an image converter tube in which each successive image is switched to a different area of the tube, so obtaining a short sequence of images with known delays. For example, each image can have an exposure of 100 ns, with an interval of 10 μs between exposures. Small, lightweight, battery-operated flash X–ray equipment has been used for conventional radiography on site work where conventional generators and power supplies are not available. The exposure is built up from a series of flashes, as the batteries continuously recharge the capacitors in the power supply. Generally, however, such small flash X–ray units have insufficient output to permit the use of fine-grain film and high quality techniques. Flash X–ray diffraction equipment is also available, as well as 'flash' electron beams.

3.11.6 Other uses of X– and gamma-rays

There are many other applications of X–rays, which by convention are usually regarded as outside the field of non–destructive testing, the most important being as follows.

X-ray diffraction

X–rays are diffracted by crystals in a manner dependent on the wavelength and the crystal lattice structure, so that X–ray diffraction provides a means for studying crystalline substances. Laue patterns can be obtained from large crystals, or Debye-Scherrer ring patterns from powder specimens, either by transmission or by reflection.

X-ray fluorescence analysis

X–ray fluorescence can be used for analysis. The emission spectrum from a specimen is recorded either by using a crystal to separate the wavelengths, or by using a proportional counter. The latter can be a lithium-drifted silicon detector with a high resolution. The main uses of X–ray fluorescence techniques are for materials with atomic number greater than 12, and the analysis can be quantitative.

X-ray stress measurement

X–rays can also be used to measure residual stress on the surface of a material by measuring the deformation of the lattice plane spacings. The effect appears as a broadening of the appropriate diffraction line. Automated equipment is available for industrial use.

Radiation gauging

A large number of radioisotopes are used for thickness gauging, both by measuring through-transmission and for beta back scatter methods. The quantity of electron back scatter is dependent on the atomic number of the specimen and the electron energy.

Autoradiography

Autoradiography is a method of producing an image of the distribution of radioactivity in a specimen on a recording medium such as a film. The film and the specimen surface are pressed into close contact for an appropriate exposure-time and the film is then removed and processed. The main applications of autoradiography are to nuclear reactor fuel elements to locate the various components, but the method is also sometimes used on specimens which have been activated in an atomic pile.

3.12   RADIATION SAFETY

Ionising radiation can cause damage to living tissue, so it is necessary to ensure that personnel operating, or in the vicinity of, radiographic equipment are adequately protected from radiation.

The subject is complicated in that it is necessary to consider both radiation-induced damage in the individual and genetic damage to part of the total population. It is also necessary to take into account that the background level of ionising radiation is not zero: there is a measurable intensity of radiation to which the whole population is exposed arising from cosmic rays (i.e. from high-energy atomic particles entering the earth's atmosphere), there are naturally-occurring radioactive elements and isotopes, and in some regions the local rock, e.g. granite, causes a higher-than-average natural radiation background. In London, the background radiation level is about 2.5 Sv per year: in some places it is much higher. Some industrial processes release radioactivity, as does atomic bomb testing, and also most humans have been subject at some stages of their life to diagnostic radiography for medical purposes. All these factors must be taken into account in determining the acceptable dose-levels around radiographic equipment.

Over the years the International Commission on Radiation Protection (ICRP) has issued a series of recommendations and revisions, which forms the basis for radiation safety requirements and calculations. Various national bodies have implemented these and added shielding data, guidelines, codes of practice, etc. In 1987 the National Radiation Protection Board recognised that the levels of risk associated with ionising radiation exposure had been underestimated, and new recommendations were issued. For details the following documents can be consulted.

(1)   *Ionising Radiations Regulations 1985*, No. 1333, HMSO, London, 1985.
(2)   *The Protection of Persons Against Ionising Radiation Arising from any Work Activity*, Approved Code of Practice, HMSO, London, 1985.
(3)   BS: 4094, *Data on Shielding from Ionising Radiations*, Part 1: 1966, *Gamma-radiation*, Part 2: 1971, *X-radiation*, British Standards Institution, London.
(4)   *Basic Radiation Protection Criteria*, NCRP Report No. 39, 1971.
(5)   *Instrumentation and Monitoring Methods*, NCRP Report No. 57, 1978.
(6)   *Protection against Neutron Radiation*, NCRP Report No. 38, 1971.

(7) *General Safety Standards for Installations up to 10 MeV*, NBS Handbook 114, US Dept of Commerce, 1975.

(8) *Standards for Protection Against Radiation*, US Govt Printing Office, 1971.

(9) *Regulations for Safe Transport of Radioactive Materials*, IAEA, 1973.

(10) *ICRP Recommendations*, 1975, Pergamon Press.

(11) BS:5650:1978, *Specification for Apparatus for Gamma-radiography*, British Standards Institution, London.

(12) *Code of Practice for Site Radiography*, Kluwer-Harrap Handbooks, 1975 (revised 1985).

References (1) and (2) are now the definitive documents for the UK with detailed shielding data in reference (3).

In the UK, the Approved Code of Practice 1985 requires that shielding must be provided around any radiographic installation so that the instantaneous dose-rate outside the shielding is as low as reasonably practical, and a value of $2.5 \, \mu Sv \, h^{-1}$ is taken as the design criterion ($0.25 \, mR \, h^{-1}$ in non-SI units). The Code of Practice should be consulted for further details.

The basis of all radiation protection standards up to 1985 was that the dose-limits for the whole body shall not be greater than:

for employees aged 18 years or over	50 mSv (5 rems)
for trainees under 18 years	15 mSv
for any other person	5 mSv

in any calendar year. Separate limits are given for dose-limits for individual organs and tissues, for the lens of the eye, and for women of child-bearing age. In 1987 it was recommended that 50 mSv be reduced to 15 mSv.

The ultimate responsibility for providing protective measures lies with the employer, and most will need to appoint a Radiation Protection Advisor with suitable qualifications and experience. Where work with ionising radiation sources makes it likely that the radiation dose may exceed a prescribed limit, 'controlled areas' must be set up with limited access. Classified personnel with access to controlled areas must use personal monitoring dosemeters. These may be direct-reading (pocket ionisation chambers), film, or thermoluminescent dosemeter badges. Film badges depend on measurement of the density on the exposed and processed film, using a calibration curve. TLDs have a sheet of thermoluminescent material which, when heated, emits light proportional to the quantity of ionising radiation absorbed. Neutron-sensitive dosemeters are also available.

The general principles of radiation protection design are:

a. distance: personnel kept away from radiation sources: the inverse square law applied;

b. absorption: barriers of highly absorbent material provided;

c. time: planned so that personnel are within a radiation field for the minimum practicable time.

Overall, the principle of ALARP—As low as reasonably possible—should always be applied to minimise radiation dosage to personnel.

Further reading

Garratt D A and Bracher D A (eds), *Real Time Radiologic Imaging*, ASTM-STP-716, ASTM, Philadelphia, USA, 1980

Halmshaw R, *Industrial Radiology: Theory and Practice*, Applied Science, London and New Jersey, 1982

Herz R H, *The Photographic Action of Ionising Radiation*, Wiley Interscience, New York and London, 1969

Huang H K, *Elements of Digital Imaging*, Prentice-Hall, New York, 1987

Industrial Radiography, Agfa–Gevaert Handbook (revised edition), Mortsel, Belgium, 1954

Mees C E K, *Theory of the Photographic Process* (revised edition), Macmillan, New York, 1954

Moores B M, Parker R P and Pullen B R (eds), *Physical Aspects of Medical Imaging*, Wiley Interscience, 1981

Non-destructive Testing Handbook, Volume 3, American Society of NDT, Columbus, Ohio, 1985

References

BS:2600:1983, *Radiographic examination of fusion butt-welded joints in steel*, British Standards Institution, London

BS:3971:1985, *Specification for image quality indicators for industrial radiography, including guidance on their use*, BSI, London

BS:2910:1986, *Methods for radiographic examination of fusion welded circumferential butt joints in steel pipes*, BSI, London

BS:6932:1988, *Measurement of the effective focal spot size of mini-focus and micro-focus X-ray tubes used for industrial radiography*, BSI, London

BS:7009:1988, *Application of real-time radiography to weld inspection*, BSI, London

Halmshaw R, *J Photo Sci*, **3**, 161, 1955

Hounsfield G N, *Phil Trans Roy Soc London*, **292(A)**, 233, 1979

Klasens H R, *Philips Res Rep*, **1**, 2, 1946

Rose A, *J Opt Soc Amer*, **38**, 196, 1948

Sturm R E and Morgan R H, *Amer J Roentgenology*, **62**, 617, 1949

4

Ultrasonic testing of materials

4.1 BASIC PRINCIPLES

Mechanical vibrations can be propagated in solids, liquids and gases. The actual particles of matter vibrate, and if the mechanical movements of the particles have regular motion, the vibrations can be assigned a frequency in cycles per second, measured in hertz (Hz), where 1 Hz = 1 cycle per second. If this frequency is within the approximate range 10 to 20 000 Hz, the sound is audible; above about 20 000 Hz, the 'sound' waves are referred to as 'ultrasound' or 'ultrasonics'.

As an example of a practical application, if a disc of piezoelectric material is attached to a block of steel (Fig. 4.1(a)), either by cement or by a film of oil, and a high-voltage electrical pulse is applied to the piezoelectric disc, a pulse of ultrasonic energy is generated in the disc and is propagated into the steel. The velocity of one common' form of ultrasonic waves in steel is approximately 6×10^3 m s^{-1}, so if the piezoelectric disc is of a suitable thickness to generate waves of frequency 1 MHz, that is a pulse of ultrasonic waves of wavelength 6 mm. This pulse of waves travels through the metal with some spreading and some attenuation and will be reflected or scattered at any surface or internal discontinuity such as an internal flaw in the specimen. This reflected or scattered energy can be detected by a suitably-placed second piezoelectric disc on the metal surface and will generate a pulse of electrical energy in that disc. The time-interval between the transmitted and reflected pulses is a measure of the distance of the discontinuity from the surface, and the size of the return pulse can be a measure of the size of the flaw. This is the simple principle of the ultrasonic flaw detector and the ultrasonic thickness gauge. The piezoelectric discs are the 'probes' or 'transducers': sometimes it is convenient to use one transducer as both transmitter and receiver—sometimes called a 'transceiver'—by detecting the return pulse between successive input pulses. In a typical ultrasonic flaw detector the transmitted and received pulses are displayed in a scan on a timebase on an oscilloscope screen (Fig. 4.1(b)). In the thickness gauge the interval between the pulses is usually displayed digitally, after conversion to a thickness scale using the ultrasonic velocity in the material being tested.

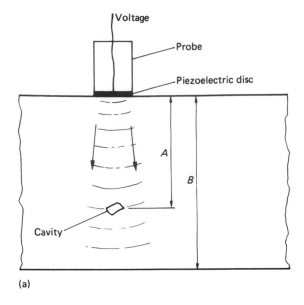

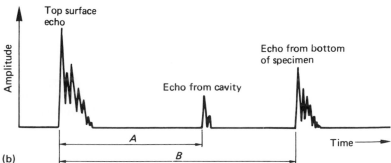

Fig. 4.1   Basic principle of ultrasonic testing with a compressional wave probe: (a) set-up of probe on specimen and (b) standard A-scan display on CRO screen

4.1.1   Waves

Ultrasonic waves are mechanical vibrations, not electromagnetic radiation, and so have a different wavelength in different materials. They are possible because of the elastic properties of the material and are due to induced particle vibration in the material. If the particle vibration is sinusoidal, the waves can be assigned a single wavelength, λ, from the well-known formula

$$f = c/\lambda \qquad (4.1)$$

c being the wave velocity. A pulse of ultrasonic energy can be considered as the synthesis of a series of purely sinusoidal waves of different frequencies and amplitudes (Fig. 4.2). The narrower the pulse, the greater the number of frequency components, as is well known from Fourier analysis.

   If the particle motion in a wave is along the line of the direction of travel of

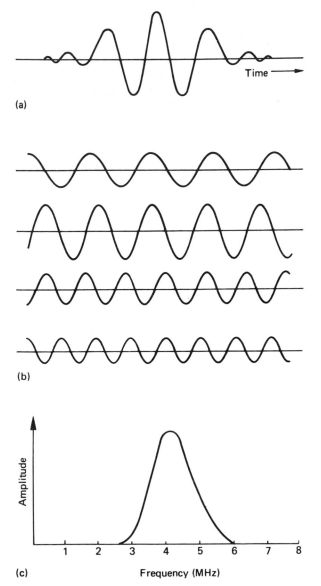

Fig. 4.2 Ultrasonic pulse: (a) pulse shape, (b) synthesis of sinusoidal waves of different frequencies and amplitudes, and (c) frequency spectrum of pulse

the wave, the resulting wave is called a *compressional wave* (sometimes also called a *longitudinal wave*) or a P-wave (primary wave). Such waves can be propagated in solids, liquids and gases.

In solid material it is possible, also, for the particle movement to be at right angles to the direction of travel of the wave—such waves are called *shear waves*. These usually have a velocity of approximately half of that of longitudinal waves in the same material and for practical purposes cannot be generated in liquids. They are sometimes called *transverse waves*. Shear waves from a typical mode-conversion transducer as shown in Section 4.3 (Fig. 4.23) are vertically polarised, but horizontally polarised shear waves are also possible.

Various different types of *surface wave* can also be produced in solid material.

Rayleigh waves are one type of surface wave which can be generated on the free surface of any solid material. They are somewhat analogous to water waves in which the motion of the particles is both transverse and longitudinal in a plane containing the direction of propagation and the normal to the surface. In Rayleigh waves the particle movement is elliptical and such waves exist only in the surface layer of large solids. At greater depths than two wavelengths below the surface the particle motion is practically zero. The surface velocity is usually about 0.9 times the longitudinal wave velocity in most solids. Rayleigh waves on a solid surface are non-dispersive, but in a layered medium dispersive Rayleigh waves (i.e. the wave velocity is frequency-dependent) are possible.

At a liquid/solid interface, where the density of the liquid is low, leaky Rayleigh waves, where there is some transfer of ultrasonic energy back into the liquid, are possible.

At an interface, either liquid/solid or solid/solid, there can also be an interface wave which is undamped. The particle displacement is in the sagittal plane. This is known as the Stoneley wave. In a thin film (solid/solid) when the particle oscillation is in a plane parallel to the interface, the waves are known as Love waves. Love waves are dispersive. In plate material where the thickness is of the order of a few wavelengths, various forms of plate wave are possible, the most important forms for NDT being Lamb waves. These are a combination of compressional and shear waves, the proportion depending on the frequency. In Lamb waves there is therefore a component of particle oscillation at right-angles to the surface and they may be regarded as complex resonances of the plate. Lamb grouped the infinite number of modes in which a plate can vibrate into two main types according to the direction of the particle displacement—symmetrical and asymmetrical. The wave velocity depends on the plate thickness, the frequency, the mode order and the material. There is a difference between the phase velocity (the rate of travel of a wave crest) and the group velocity (the velocity of a short pulse of waves).

Lamb waves are dispersive. Flexural waves, or guided waves, similar to Lamb waves can be produced in wires (rod waves) or tubes. Rayleigh, Lamb and Love waves are used for special applications in ultrasonic testing of materials, but by far the most important types of wave for industrial ultrasonic applications are the compressional and shear waves.

It should be remembered that plane waves are obtained only with large plane sources applied to large specimens. With small sources, the propagation of waves in plates differs considerably from the ideal concept of the Lamb wave.

The terms 'creeping wave' or 'longitudinal creeping wave' or 'longitudinal head wave' have been used for a kind of surface/near surface wave, very similar to a Rayleigh wave, which can be excited at a metal surface. The wave 'creeps' along the surface, but otherwise exhibits the properties of a compressional wave travelling at 90° to the normal to the surface. The wave intensity is low, but is not affected by surface irregularities. The use of these names through the NDT literature is not completely consistent.

Special forms of interface wave at solid/solid interfaces are important for the evaluation of adhesively-bonded joints.

4.1.2   Wave velocity

The velocities of the various kinds of ultrasonic wave can be calculated from the elastic constants of the material. For compressional waves, in a specimen of large dimensions compared to the wavelength,

$$V_C = \left[\frac{E(1-\sigma)}{\rho(1+\sigma)(1-2\sigma)} \right]^{1/2} \tag{4.2}$$

The shear wave velocity is given by

$$V_S = \left[\frac{E}{2\rho(1+\sigma)} \right]^{1/2} \tag{4.3}$$

where E is Young's modulus (Nm^{-2}), V_C is the compressional wave velocity, V_S is the shear wave velocity, ρ is the density of the specimen, and σ is Poisson's ratio.

Table 4.1 gives ultrasonic velocities in a selection of materials. The velocity of Lamb waves cannot be specified as it depends also on the plate thickness and the frequency. In addition, for a pulse source there is a dispersion of the speed of propagation. Even with compressional and shear waves there are small deviations from the values given in Table 4.1 for crystalline materials, because of elastic anisotropy: this is important and particularly marked in copper, brass and austenitic steels.

Table 4.1   Ultrasonic velocities (mean values)

Material	Relative density	Velocity (m s^{-1})	
		Compressional	Shear
Aluminium	2.70	6300	3080
Mild steel	7.85	5900	3230
Magnesium	1.70	5770	3050
Copper	8.90	4700	2260
Titanium	4.51	6000	3000
Polythene	1.20	2000	540
Perspex (Lucite)	1.18	2700	1300
Water	1.00	1490	—
Air	—	344	—

4.1.3   Standing waves

If two waves are generated at two different points in a specimen, and have the same frequency and amplitude but different directions of propagation, these waves can interfere and can result in a standing wave. For example, if particles in a material have exactly opposite directions of propagation, they can remain constantly at rest (Fig. 4.3). This form of wave is well known in the oscillations of a taut string and occurs frequently in ultrasonic testing during the reflection of a wave from a smooth surface: such standing waves can cause considerable confusion with continuous wave systems, but are less of a problem with the more-widely-used pulse wave techniques.

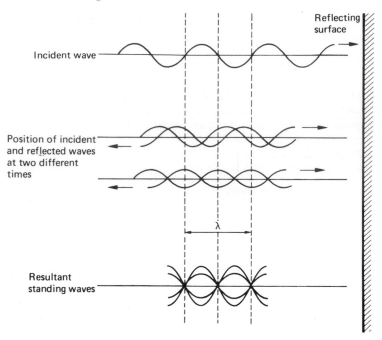

Fig. 4.3   Formation of standing waves, due to mode conversion at a plane reflector

4.2   WAVES AT BOUNDARIES

4.2.1   Acoustic impedance

When an ultrasonic wave is incident on a plane boundary between two media, perpendicular to the surface, some ultrasonic energy is transmitted through the boundary and some is reflected. The percentages of energy transmitted and reflected depend on the *specific acoustic impedance*, Z, defined for each material as

$$Z = \rho V \tag{4.4}$$

where ρ is the density of the material and V is the velocity of the wave.

For two materials of different acoustic impedances, Z_1 and Z_2, the percentage of energy transmitted, E_T, is given by

$$E_T = \frac{4 Z_1 Z_2}{(Z_1 + Z_2)^2} \times 100 \tag{4.5}$$

and the reflected energy E_R, by

$$E_R = \left(\frac{Z_1 - Z_2}{Z_1 + Z_2}\right)^2 \times 100 \tag{4.6}$$

These formulae are valid for both compressional and transverse waves, but as a transverse wave cannot be sustained in a liquid, a transverse wave is always completely reflected at a solid/liquid or solid/gas interface.

A common practical case is the water/steel (or steel/water) interface. Inserting suitable values, it can be calculated that at a water/steel interface, 12% of the incident energy is reflected and 88% is transmitted.

It should be noted that formulae 4.5 and 4.6 are for transmitted and reflected *energies*; for amplitude values, the square root is taken. These formulae are for single, large-area interfaces, but the double interface is also of practical importance (Fig. 4.4).

The wave in material A is split at the interface between A and B into a transmitted and reflected wave and the transmitted component is again divided at the interface between B and C and so on: the result is a sequence of reflected waves in both directions between A and C, and depending on the wave phases there may be interference in both the reflected and transmitted components. Maximum transmission occurs when the distance *d* is an integral number of half-wavelengths and minimum transmission when *d* is an odd number of quarter-wavelengths. The effect is of importance in determining the thickness of liquid couplant used as an interface between the piezoelectric element of a probe and the specimen surface. For optimum transmission, the couplant should have a thickness of one half-wavelength (or a multiple thereof).

A second special case of the multiple interface is an air-filled crack in metal with a very narrow (much less than one wavelength) opening width. Calculated results given by Krautkrämer show that with a gap of about 10^{-6} mm in a steel specimen, the calculated theoretical reflection from the crack is about 70%, and larger gaps reflect effectively 100%. Only therefore with extremely tight cracks is there a possibility of partial transmission across an air gap. In practice, because of the irregularities in 'real' crack opening widths and the influence of foreign material on the crack surfaces, apparently wider cracks can be semi-transparent to ultrasonic energy: nevertheless, unless

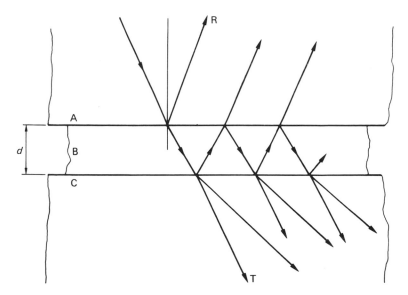

Fig. 4.4 Ultrasonic wave on an interface between two materials, A and C, with a coupling layer, B. T denotes the transmitted beam and R the reflected beam

the crack opening width is less than 1 μm, there should be no practical problem in having sufficient reflected ultrasonic energy for crack detection.

4.2.2   Waves at oblique incidence

When an ultrasonic beam is incident at any angle except the normal at an interface between two media having different acoustic impedances, it can produce both reflected and refracted compressional and shear waves.

A simple relationship, known as *Snell's law*, describes the angle of refraction (Fig. 4.5) of the transmitted wave,

$$\frac{\sin \alpha}{\sin \beta} = \frac{V_A}{V_B} \tag{4.7}$$

where α and β are the angles of incidence and refraction respectively and V_A and V_B are the wave velocities in the two media A and B.

For the reflected wave in medium A, the angle of incidence is equal to the angle of reflection. These expressions hold for both incident compressional and shear waves. When $V_B > V_A$ it is possible to have an angle of incidence α which would make $\beta = 90°$. α is then referred to as the *critical angle*, and for angles of incidence greater than this the wave is totally reflected and no energy is transmitted into the second medium. In the case of a water/steel interface, the critical angle for a compressional wave is about 15°. At the interface between two solid media there are two critical angles, one at which the transmitted compressional wave disappears and one beyond which the transmitted shear wave no longer exists. At a Perspex/aluminium interface, such as

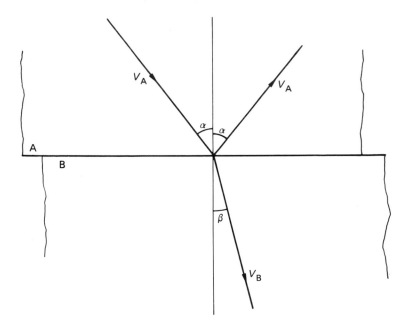

Fig. 4.5   Ultrasonic wave at an angle on an interface between two materials, A and B, in which the waves have different velocities, with $V_A > V_B$

that in a shear wave probe, these angles are 25.5° and 61.3° respectively.

At an interface it is possible to have wave mode conversion, and Figs 4.6 and 4.7 show the general cases of an incident compressional and shear wave respectively.

In Fig. 4.6,

$$\frac{\sin i}{V_{CA}} = \frac{\sin r_S}{V_{SA}} = \frac{\sin r_C}{V_{CA}} = \frac{\sin R_S}{V_{SB}} = \frac{\sin R_C}{V_{CB}}$$

and in Fig. 4.7,

$$\frac{\sin i}{V_{SA}} = \frac{\sin r_C}{V_{CA}} = \frac{\sin R_S}{V_{SB}} = \frac{\sin R_C}{V_{CB}}$$

where V_{CA} and V_{SA} are the velocities of the compressional and shear waves respectively in medium A and V_{CB} and V_{SB} are the velocities of the compressional and shear waves respectively in medium B.

In both cases, depending on the incident angle, some of the secondary waves may not exist.

It is also important to know the relative amplitudes of all the waves that can exist during mode conversions and the equations for these were worked out originally by Knott (1899) in connection with seismic waves, with modifications by later authors. The relationships are very complex. Figure 4.8 shows the relative amplitudes calculated for a Perspex/aluminium interface, for an incident compressional wave at different angles. There is a sharp peak in the amplitude of the reflected compressional wave at the first critical angle (25.5°), and another in the reflected shear wave at the second critical angle (58°) when the transmitted shear wave disappears. It is the region between 0° and 25.5° which is most important if the transmitted compressional wave is to be used for flaw detection.

A simpler case is the steel/air interface, where there is no transmitted ultrasonic energy. Two cases, for incident compressional and shear waves respectively, are shown in Figs 4.9 and 4.10. These show the relative amplitudes of the reflected compressional and shear waves.

In practical ultrasonic testing, certain cases are particularly important. The solid/solid case occurs with contact probes on metal surfaces, although usually a thin layer of liquid couplant is used between the solids and this liquid cannot transmit shear waves, so the practical case is solid/liquid/solid. For shear wave inspection, which is widely used in weld inspection, the incident angles of interest are those between the two critical angles and the usual requirement is for a transmitted shear wave at 45°–80°. If a 70° shear wave beam (i.e. 70° to the normal) is required in steel, then the angle of incidence in Perspex of the incident compressional wave can easily be calculated by Snell's law to be 54°.

The water/metal case occurs with stand-off probes and with immersion testing. The efficiency of energy transmission across the interface varies markedly with angle. These show that for incident angles up to 30° it is better to operate with compressional waves, but above 35° shear waves become more favourable.

In this discussion of mode conversion, the shear wave has been considered to have its plane of oscillation in the plane of incidence, and only this sort of shear wave is produced when there is mode conversion at an interface. An

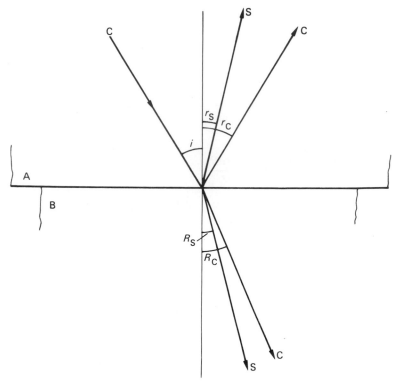

Fig. 4.6 Compressional wave at an angle on to an interface between two materials, A and B, showing mode conversions. C denotes compressional waves and S shear waves, $r_C = i$ (the angle of incidence), and $V_A > V_B$

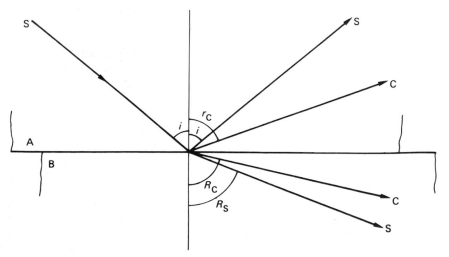

Fig. 4.7 Shear wave at an angle on to an interface between two materials, A and B, showing mode conversions. C denotes compressional waves and S shear waves, and $V_B > V_A$

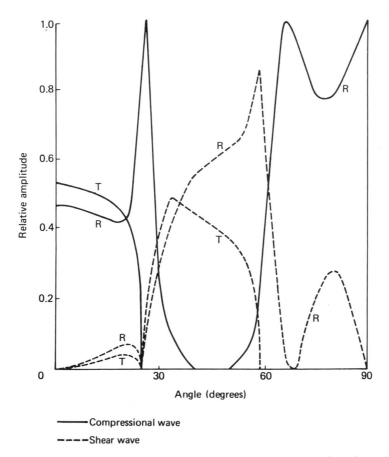

Compressional wave

Shear wave

Fig. 4.8 Relative amplitudes, calculated for a Perspex/aluminium interface, for an incident compressional wave at different angles. T denotes transmitted waves and R reflected waves

incident shear wave having a plane of oscillation at right angles to the plane of incidence will be totally reflected at all angles and will not divide into another type of wave. During mode conversion the reflected wave has its phase reversed (i.e. shifted half a wavelength), except in the case of an incident shear wave polarised at right angles to the plane of incidence.

In practical ultrasonic testing, an ultrasonic beam is frequently incident on a right angled edge or corner, and as there are two surfaces involved there can be complex mode conversions and energy changes (Fig. 4.11). With a compressional wave, except for very flat angles, there is an apparently very bad reflection of energy, the reason being that at one of the two surfaces much of the energy goes into a strong shear wave which does not return to the direction of the incident wave. With an incident shear wave the intensity of reflection depends markedly on angle. In steel, there is almost zero reflection at 30° and 60°, with strong reflection between these angles and poorer reflection outside these angles. This point has considerable significance in flaw detection with shear waves, where a beam incident on a planar flaw at 30° or 60° may produce

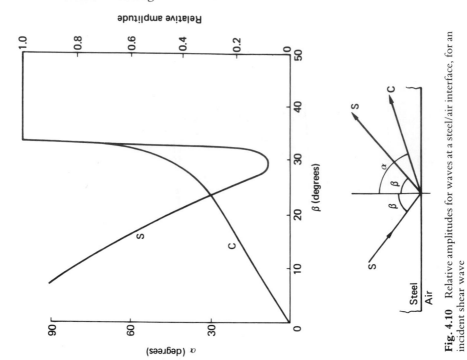

Fig. 4.10 Relative amplitudes for waves at a steel/air interface, for an incident shear wave

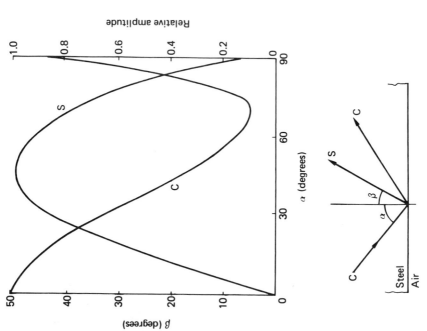

Fig. 4.9 Relative amplitudes for waves at a steel/air interface, for an incident compressional wave

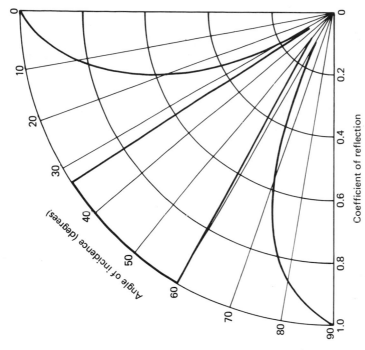

Fig. 4.12 Coefficient of reflection for an incident shear wave at different angles, in steel

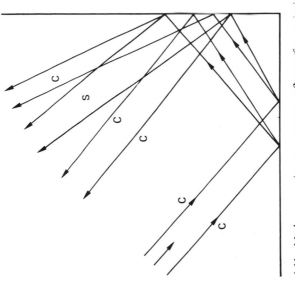

Fig. 4.11 Mode conversion at a corner reflector for an incident compressional wave

a very poor indication. A calculated reflection polar diagram is shown in Fig. 4.12.

4.2.3   Surface waves

If a compressional wave strikes an interface at approximately the critical angle for total reflection of the shear wave, a surface (Rayleigh) wave is produced along the interface. For a Perspex/steel interface this angle is 57.8° in Perspex (Lucite). This surface wave is strongly attenuated over the length of the Perspex, but is unattenuated thereafter, unless affected by the roughness of the surface or surface-breaking flaws. The greatest intensity of surface wave is produced at an angle of incidence slightly greater than the critical angle.

True plate waves with a direction of oscillation parallel to the surface of the plate are not produced by mode conversion, but by specially–designed probes. Lamb waves, however, can be produced by having an incident compressional wave of suitable frequency and angle of incidence on a plate specimen. The correct angle of incidence, θ, is given by

$$\sin \theta = V_\mathrm{L}/V_\mathrm{P}$$

where V_L is the velocity of the incident wave and V_P is the phase velocity of the desired Lamb wave. It is necessary to have a calibration graph of phase velocity against (frequency × plate thickness) (Fig. 4.13). The phase velocity is the velocity with which individual Lamb waves travel along the plate and first mode Lamb waves have a similar velocity to shear waves. The vertical portions of the curves of Fig. 4.13 represent regions where the various modes

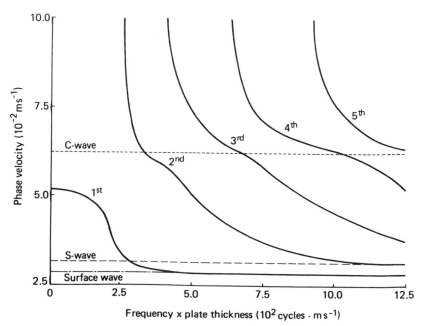

Fig. 4.13 Lamb waves: phase velocity/(frequency × plate thickness) for different modes of symmetrical waves

propagate dispersively while in the horizontal regions there is very little attenuation. Both symmetrical and asymmetrical Lamb waves are possible. When Lamb waves are reflected at an edge they can change part of their energy into another mode or type.

4.2.4 Ultrasonic optics

Geometric optics uses light rays which can be drawn as straight lines and this method has already been used with ultrasonic beams to describe simple reflection and refraction. It must be rememebred, however, that the wavelength of light is of the order of 10^{-3} mm, whereas the wavelength of ultrasonic waves is of the order of 1 mm. For ultrasonic waves, therefore, diffraction phenomena are likely to occur more frequently and to be more important, particularly as in practice ultrasonic beams are used at ranges from a few wavelengths, up to perhaps a few hundred wavelengths. Lenses and mirrors can be used with ultrasonic waves, but the simple geometric optics treatment will give only a rough approximation to the practical result.

If an ultrasonic wave is considered to have a spherical wavefront, the intensity varies inversely with the (distance)^2, the well-known *inverse square law*. This is not true for a plane wavefront, nor for a cylindrical wavefront. In the latter case, the intensity changes only inversely with the distance.

If a spherical wavefront wave is reflected at a plane surface, a simple geometric construction of the wavefront shows that the shape of the spherical wavefront is preserved and the beam also retains its beam width. As described in Section 4.2.2, however, the reflective coefficient depends strongly on the angle of reflection. If there is mode conversion, the refracted beam may no longer be spherical.

If the surface between two media with different acoustic impedances is not plane, the ultrasonic beam may be focussed or made divergent, similar to the effects produced by curved mirrors with light, and a focal length, F, can be assigned to a spherical interface.

For the reflected beam, the simple optical formulae can be used:

for a plane wave $\qquad F = \dfrac{R}{2}$

for a spherical wave $\qquad \dfrac{1}{b} \pm \dfrac{1}{a} = \dfrac{1}{F}$

whereas for the transmitted beam (Fig. 4.14):

for a narrow beam $\qquad F = \dfrac{R}{(1 - V_2/V_1)}$

for a spherical beam $\qquad \dfrac{1}{b} - \dfrac{V_2/V_1}{a} = \dfrac{1}{F}$

where F is the focal length, R is the surface radius, a is the wavefront radius, b is the image front radius, and V_1 and V_2 are acoustic velocities in the two materials.

The ratio V_2/V_1 is analogous to the refractive index with light and, as at a

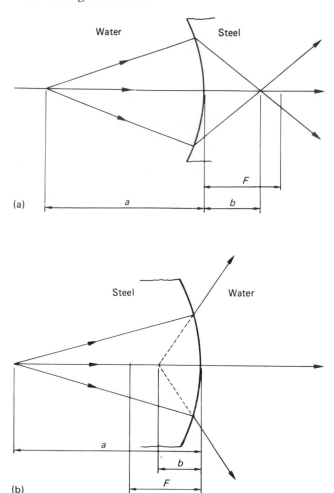

Fig. 4.14 Ultrasonic waves incident on (a) curved water/steel and (b) curved steel/water interfaces

water/steel interface, it will have a value of about 4, this produces a very strong refracting effect. It is also one of the reasons for difficulties encountered in testing objects with a curved surface by immersion techniques: unwanted mirror and lens effects are produced by the curved surfaces.

Acoustic lenses, both spherical and parabolic, are used with focussed probes (see Section 4.6). To maintain high efficiency in the case of ultrasonic lenses, the reflection losses should be kept small, which means that the acoustic impedance of the lens material and adjoining material should be as nearly as possible equal, but for satisfactory refraction a large difference in ultrasonic velocity is necessary. These are, to a large extent, contradictory requirements and would seem to rule out metal lenses immersed in water. A compromise, and practical answer, is an aluminium lens, faced with an acrylic reverse lens with an outer flat face (Fig. 4.15).

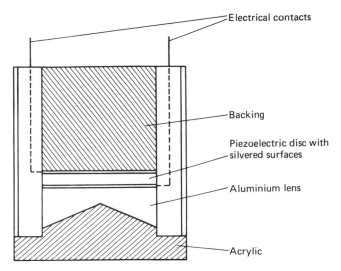

Fig. 4.15   Focussed beam ultrasonic probe

4.2.5   Diffraction effects

If a plane wavefront is beamed through a slit in a diaphragm, the sound field is changed from a sharply defined cylinder by diffraction effects. Huyghens' principle would give a plane wave, but to this must be added edge effects. In the case of a circular diaphragm aperture, an annular wave is produced which, on superposition on the plane wavefront, produces a field of maxima and minima as shown diagrammatically in Fig. 4.16. This is an instantaneous condition and the maxima travel with the wave along the dotted curves. The ratio of the aperture (of the ultrasonic source) to the wavelength determines the spread of the ultrasonic field and the number of maxima and minima. Figure 4.16 is over-simplified, and actual fields must be computed three-dimensionally: an example of the sound field in front of a plane source, with a 10 mm diameter probe of 5MHz frequency, in water, is shown in Fig. 4.17. If the source is considered as a simple piston source, the variation in acoustic pressure, P_R, along the beam axis at a distance R is given by

$$P_R = P_0\, 2\sin\left\{\frac{\pi}{\lambda}\left[\left(\frac{D^2}{4}+R^2\right)^{1/2}-R\right]\right\}$$

Along the axis of the beam, the last maximum is located at a distance N from the source, where

$$N = (D^2 - \lambda^2)/4\lambda$$

which can usually be taken as approximately equal to

$$N = D^2/4\lambda$$

Distance N is known as the *near-field length* and the ultrasonic field beyond N is called the *far field*. The terms 'Fresnel region' and 'Fraunhöfer region' are sometimes used for the near and far fields respectively.

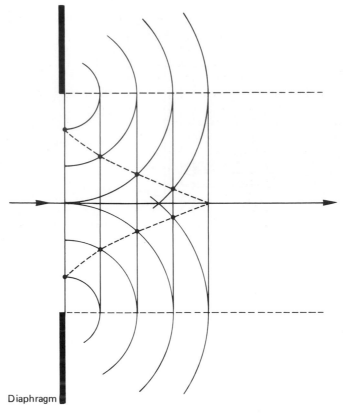

Fig. 4.16 Interference structure of the sound field, calculated according to Huyghens' principle, for a circular aperture

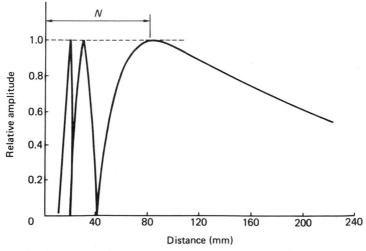

Fig. 4.17 Sound field along the axis of a 10 mm, 5 MHz probe in water, where N is the near-field length

In a typical ultrasonic flaw detection technique, the probe diameter may be 20 mm, and, in steel, with 2 MHz frequency, the wavelength will be 3 mm, so the near-field length with be 33.3 mm. In water, with the same probe, the near-field length would be 134 mm. If a reflecting flaw is located within the near field of a probe, obviously the intensity of the reflected energy will depend on whether the flaw is on a maximum or a minimum intensity point in the beam and it will not be possible to make direct use of the intensity of the reflected energy as a measure of the cross-sectional area of the flaw.

Based on a plane, uniformly-emitting, ultrasonic source, the profile of an ultrasonic beam is as shown diagrammatically in Fig. 4.18. This is very much a theoretical beam shape, as in practice the beam edges are not sharp cutoffs and edge effects have been neglected; furthermore, this presupposes a continuous wave rather than the ultrasonic pulse which is normally used in ultrasonic flaw detection.

If the beam spread is measured as angle θ (Fig. 4.18), diffraction theory gives the relation

$$\sin(\theta/2) = 1.2\lambda/D$$

the beam width to half the centreline intensity by

$$\sin(\theta/2) = 0.56\lambda/D$$

and the width to one-tenth of the centreline intensity by

$$\sin(\theta/2) = 1.08\lambda/D$$

These equations for beam spread are only valid for small values of λ/D and therefore become increasingly in error with smaller ultrasonic sources, where the angle of divergence of the beam is approximately 90°. In the far field, the beam intensity along the beam axis follows the inverse square law, neglecting absorption and scattering effects.

Practical ultrasonic probes are usually far from being uniformly-emitting sources, and if the ultrasonic intensity distribution along the field of a probe is required, it is essential to determine this experimentally for each particular probe; many laboratories are equipped to do this as part of a specification of the probe characteristics. The measurement is usually made with a point detector in a water tank (see also Section 4.4.2).

It should be noted that the equations for beam spread show that a small diameter probe will have a wide beam, and a large diameter probe a more directional beam (Fig. 4.19), but with a longer near field.

There is one other effect of the ultrasonic field which needs to be considered: this is the effect produced at a flaw in a specimen when the flaw has dimensions of the same magnitude as the ultrasound wavelength.

First of all, a considerable amount of energy is diffracted round the flaw into the 'shadow space', where there will be maxima and minima of intensity. Secondly, if the flaw is small, every point on the flaw becomes a source of reflected elementary waves, and the flaw can be regarded as a secondary source of ultrasound with its own near and far fields. Therefore, for a small flaw, the height of the returning echo from the flaw is directly proportional to the cross-sectional area of the flaw and inversely proportional to the distance from the flaw to the receiver. For a large reflector, such as the backwall of a

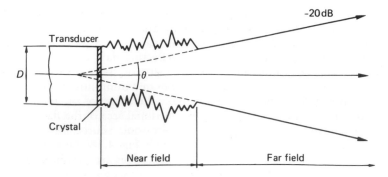

Fig. 4.18 Diagrammetric representation of the sound field along the axis of a probe

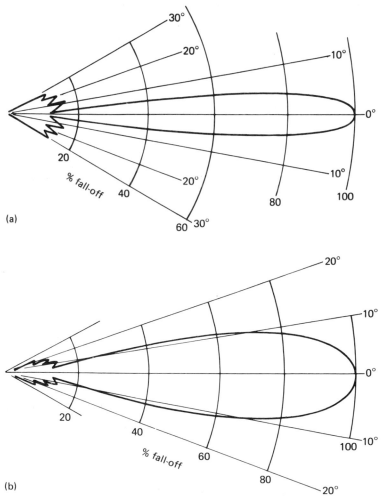

Fig. 4.19 Polar diagrams of the sound fields of two probes of different diameters: (a) $D = 24$ mm and $D/\lambda = 16$, and (b) $D = 6$ mm and $D/\lambda = 6$

specimen, the reflected sound beam intensity is inversely proportional to the distance of the wall. These relationships lead to the production of DGS (distance–gain–size) diagrams which are used extensively in defect-size determinations (see Section 4.4).

In recent years, considerable effort has been put into mathematical modelling, to predict the response of different shapes of defect to an incident ultrasonic wavefront. Besides the usual geometrical optics of reflection, refraction and transmission already described, the diffraction occurring at boundaries such as crack edges can be treated by the geometrical theory of diffraction first propounded by Keller (1957, 1962). This theory has been applied by Temple and others to the time-of-flight diffraction (TOFD) signals—see Section 4.5.6—which can be used to measure crack height, and the computer program developed for this can be adapted to cover the usual pulse echo techniques and predict defect responses. The theory is complex, is still under development, and is too long to be conveniently covered in this chapter.

4.2.6   Attenuation

In most materials, the ultrasonic intensity is reduced as the ultrasonic beam travels through a material due to various mechanisms, including scattering. The absorption depends markedly on the nature and structure of the material (grain size and grain orientation) and is also a function of the ultrasonic frequency. Many materials are anisotropic so far as absorption is concerned: that is, the absorption varies with the beam direction. Formally, ultrasonic attenuation is described in terms of an attenuation coefficient, α:

$$I = I_0 \exp(-\alpha t)$$

where I is the intensity at a distance t from an initial intensity I_0. Generally, α is taken as the sum of the true attenuation coefficient α_T and the scatter coefficient α_s.

A convenient way of measuring changes in intensity, or amplification, is in terms of decibels (dB). A decibel is one-tenth of a bel, which is a unit based on logarithms to base 10, so that if the two powers are P_1 and P_2 they are said to differ by n bels if

$$P_1/P_2 = 10^n$$

or $\qquad n = \log_{10}(P_1/P_2)$ bels

$$= 10 \log_{10}(P_1/P_2) \text{ decibels (dB)}$$

Acoustic power (or intensity) is proportional to (amplitude)^2, so for comparison of amplitudes A_1 and A_2 the equation can be rewritten

$$n = 10 \log_{10}\left(\frac{A_1^2}{A_2^2}\right) = 20 \log_{10}\left(\frac{A_1}{A_2}\right) \text{ decibels}$$

Thus an amplitude ratio between two waves of 2:1 corresponds very nearly to 6 dB.

Decibel values can be added arithmetically; amplitudes must be multiplied.

Decibel value	Amplitude ratio
3	1.41
6	2.00
10	3.16
20	10.00
30	31.60
80	10 000.00

Frequently, ultrasonic attenuation coefficients are given in dB mm^{-1} and for many engineering materials such as mild steel are very low (e.g. 5×10^{-3} dB mm^{-1} at 2 MHz).

In ultrasonic flaw detection, an attenuator circuit (sometimes called a 'calibrated gain control') is a normal part of modern equipment, and this is also calibrated in decibels. Signals are compared by using the attenuator to bring each signal to a predetermined height on the screen display and subtracting the two decibel readings obtained. Thus, if one signal requires 8 dB to normalise it and another 18 dB, the second signal has a 3.16 times greater amplitude. That is, if the first signal gave a display 31.6 mm high on the screen, the second signal height would be 100 mm, without any added attenuation.

As stated earlier, ultrasonic attenuation is also a function of frequency and the relationships are extremely complex. According to Mason and McSkimin, there is a true absorption which is proportional to frequency and then a scatter coefficient which is strongly dependent on the D/λ ratio (see also Section 4.7.5).

The attenuation coefficient can be measured by placing a transducer on a parallel-sided specimen and observing the envelope of successive backwall

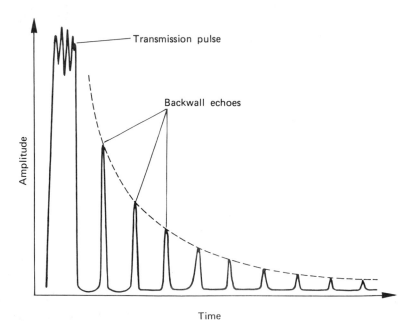

Fig. 4.20 Measurement of attenuation: multiple backwall echo method, with a parallel-sided specimen

echoes (Fig. 4.20). The specimen must be accurately plane and parallel-sided and there must be enough multiple echoes to determine the shape of the envelope. With low-attenuation materials, fluctuation in probe coupling can lead to losses which may be greater than the attenuation losses.

A particular problem is in the ultrasonic inspection of austenitic steel, particularly welds in austenitic steel, where the size, orientation, and elastic anisotropy of the different grains result in large scattering, beam distortion, a variation in ultrasonic velocity with beam direction, beam-skewing and changes in attenuation with beam orientation (see also Section 4.6.3).

Most metals show a pronounced reduction in attenuation if their cast structure is refined by cold or hot working (forging, rolling, etc.). This is because the large grains are destroyed, resulting in reduced scattering, and also the increase in the mechanical bond at the grain boundaries which is produced by working a cast structure. This effect is particularly pronounced in non-ferrous metals and high-alloy steels where even a small thickness of as-cast material cannot sometimes be penetrated by ultrasound.

4.3   GENERATION OF ULTRASONIC WAVES

There are a number of ways in which ultrasonic waves can be produced, but for non-destructive testing, most equipment uses the *piezoelectric effect*, or its inverse, for both transmitter and receiver probes. A piezoelectric material has the property that if deformed by external pressure, electric charges are produced on the surface. If the pressure is reversed, that is becomes a tension, the electric charges produced on the faces are reversed in sign. The *inverse piezoelectric effect* is that if a piece of piezoelectric material is placed between two electrodes and an electric potential is applied, the material changes shape. Thus, if the electric potential is alternating, a mechanical oscillation is produced in the piezoelectric plate. The electric potential and the mechanical pressure are directly proportional. A great number of materials have been found to be piezoelectric: both natural crystals such as quartz and lithium sulphate, and fabricated polycrystalline ceramics such as barium titanate, which can be polarised to behave like a piezoelectric material.

In natural crystals, the piezoelectric effect is a property of the crystal structure and depends on the crystal having polar axes. In quartz, there are three polar axes and if a plate is cut from the crystal at right angles to one of the axes, it exhibits particular modes of vibration. A plate cut at right angles to the x-axis is called an x-cut plate or crystal, and is the form normally used in ultrasonic probes to generate compressional waves. A y-cut plate oscillates in shear, so that if it is attached to a metal surface, a beam of shear waves is produced in the metal at right angles to the surface: this is a possible method of producing shear waves for flaw detection, but is not widely used because the piezoelectric crystal has to be cemented on to the specimen. With an x-cut crystal, while the predominant oscillations are longitudinal, there is also a shear component because of changes in the transverse dimensions when an electric potential is applied, but if the crystal is coupled to the specimen by a liquid couplant, these secondary oscillations cannot be transmitted into the specimen.

Normally, a plate of piezoelectric material is cut to vibrate at the fundamental frequency required, but plates can also be made to vibrate at harmonic frequencies, which are multiples of the fundamental frequency.

The fundamental frequency, f_0, for a quartz transducer is given by

$$t = \lambda_0/2 \quad \text{or} \quad f_0 = V/2t$$

where t is the thickness of the crystal and λ_0 is the fundamental wavelength.

So for a 4 MHz compressional wave probe, a plate 0.74 mm thick is necessary. Thus, for short wavelength probes, operating at the fundamental frequency, very thin, fragile piezoelectric plates are necessary.

A piezoelectric plate can be caused to oscillate either by an alternating voltage of suitable frequency, usually applied to metal electrodes deposited on the plate surfaces, or by a simple electrical pulse-shock excitation. In the latter case, if left to itself, the plate oscillates freely, but due to internal friction and energy radiation, the oscillations are damped and decrease rapidly in amplitude. A piezoelectric plate can also produce forced oscillations by the application of an alternating voltage of a different frequency, but the amplitude will be less than the oscillations at the natural, resonant frequency. As short, high–amplitude pulses are desirable for good ultrasonic flaw detection, the damping coefficient and quality factor (the output at resonance) of different piezoelectric materials are of considerable importance.

Besides quartz, the only other monocrystalline material used for flaw detection probes is lithium sulphate, which has a greater output than quartz and is claimed to be capable of producing shorter ultrasonic pulses than a comparable quartz probe.

In recent years, there has been much more interest in probes using a sintered ferroelectric ceramic material which can be polarised by the application of a strong electric field (1000 V cm^{-1}), applied across the thickness at the Curie temperature. This aligns one axis of the small crystals which then become 'frozen', and the plate behaves as if it is a disc of piezoelectric material. Barium titanate ($BaTiO_3$) is a polycrystalline material which can be treated in this way.

In general, the most important property for ultrasonic probes is the coupling coefficient, hence the interest in the newer ceramic sintered materials.

Physically, quartz is very hard; chemically, it is almost inert. Lithium sulphate is very sensitive, but decomposes at 75°C. Barium titanate and PZT are sintered, then shaped by grinding, and then polarised. They have a lower hardness than quartz, with a lower resistance to wear, but can be used to generate a much greater ultrasonic output. Of those listed, barium titanate and PZT are the best transmitter materials, but both have a high acoustic impedance which prevents their sensitivity being fully exploited when used in liquids or with liquid coupling to solids. There is a good theoretical case for using different piezoelectric materials for transmitter and receiver in a two-probe technique: for example, PZT or barium titanate for the transmitter and quartz for the receiver. Ceramic material probes can suffer from ageing, particularly with a probe used for immersion testing which is continually immersed in water. Most piezoelectric materials have a high Z-value and so are not well matched to either water or liquid couplant.

In practical ultrasonic testing, the need is for a short pulse, and the ideal frequency response of the piezoelectric layer, to ensure this, is a combination of a Gaussian shape to the frequency spectrum and a linear phase characteristic.

A considerable modification to the output spectrum can be produced by appropriate backing to the piezoelectric layer.

A highly damped crystal is effectively a broad-beam emitter, the shorter the pulse, the wider being the frequency spectrum. Such probes are claimed to have advantages in the examination of very thin specimens and can produce a high flaw-resolving power.

The typical design of a single crystal compressional wave probe is shown in Fig. 4.21. Such a probe is called a 0° compressional wave probe, or, in the USA, a 'straight-beam' probe. If a metal wedge is placed between the piezoelectric element and the specimen surface, of the same material as the specimen, compressional waves can be propagated into the specimen at an angle.

The piezoelectric element (the crystal or plate), of suitable thickness to produce the resonant frequency required, is usually circular in shape, and typical diameters are 6–30 mm, with frequencies in the range 1–15 MHz. The crystal faces are metallised, either by coating them with electroconductive ink which gives a deposit of silver or copper after baking, or a better coating can be obtained by vacuum deposition of nickel with an over-coating of aluminium produced by ionic bombardment. For contact probes used on metal specimens, using quartz, the front surface metallisation is unnecessary, but the probe must contain a contact spring on the specimen surface. In some probes, attempts have been made to compensate for the mechanical restrictions of the crystal mounting by using a patterned electrode coating, for example a star-shape. The aim is to obtain as nearly as possible a plain wavefront output. Alternatively, a non-metallised strip should be left around the crystal perimeter. The piezoelectric crystal is backed with a damping backing as shown in Fig. 4.21. This material must have a similar acoustic impedance to that of the crystal, so that the back wave travels into it without reflection; it should be highly absorbent, and obviously also it must be well bonded to the piezoelectric element. Electrical contacts are made by wires through this backing on to

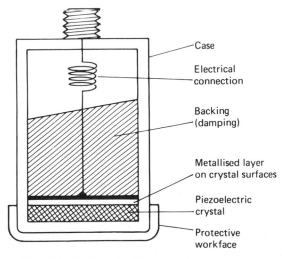

Fig. 4.21 Design of a 0° compressional wave probe

the metallised surfaces of the crystal. Nowadays, the acoustic backing is one of two kinds:

(1)   a scattering, diffusing backing, made of tungsten powder in epoxy resin or some form of sintered metal,
(2)   a quarter-wavelength layer.

Some newer piezoelectric materials such as lead niobate have a high intrinsic internal damping and can be used in probe construction without additional backing. Ceramic piezoelectric materials can also be used for constructing transducers with a curved surface, by grinding to the required curvature: the piezoelectric crystals are oriented normal to the surface at every point, during polarisation. For focussed probes, however, alternative methods of construction are usually used (see Section 4.3.3).

A new form of ultrasonic probe utilises a piezoelectric polymer material PVDF (polyvinylidene fluoride), which is available in sheet form. This is most suitable for immersion probes, as the material is acoustically well matched to water: it is highly damped, so can produce single cycle pulses, and has a very broadband frequency response. Being sheet, it is highly reproducible and so is suitable for multiprobe arrays. One practical difficulty is that the material is electrically inert. PVDF has been used to make probes operating between 0.1 and 25 MHz. PVDF is available only in very thin sheet form (9–30 μm), so its efficiency as a receiver is poor, but its transmitting efficiency is about 60% of that of PZT, so good transmitting probes can be constructed for a wide frequency range from 10 to 100 MHz. Focussing can be achieved by deforming the foil.

There are two other basic designs of ultrasonic probe—the twin crystal compressional wave probe (Fig. 4.22) and the single crystal shear wave probe (Fig. 4.23).

The former has two crystals of piezoelectric material, one transmitting and one receiving, often with Perspex 'shoes' between the crystals and the specimen. If such Perspex shoes are used, much of the near zone of the crystal can be contained within the Perspex. In Fig. 4.22, the piezoelectric crystals are both parallel to the specimen surface, but they can also be deliberately tilted inwards to provide a transceiver probe specially suitable for near-surface flaw detection. The sensitivity of such a probe can be very good for flaws at 5–10 mm below the surface, but the sensitivity falls off very rapidly for flaws which are deeper in the specimen.

The shear wave probe (Fig. 4.23) is the most-widely-used type for weld inspection. The piezoelectric element is cemented to the sloping face of a Perspex block, the angle of this face to the base being chosen so that when the Perspex flat face is placed on a metal specimen, the compressional wave in the Perspex is mode converted into a shear wave in the specimen, at a chosen angle (usually 45°, 70° or 80°). The angle of the shear wave beam, for any particular probe, will of course depend on the velocity of ultrasound in the specimen material. Thus (see Table 4.2) a 70° probe for use on steel is *not* a 70° probe when used on aluminium. Such probes are commonly sold in terms of the nominal angle of the shear wave beam in steel.

Materials other than Perspex have been proposed for the probe wedge in a shear wave probe, but the requirement for a material in which the compressional wave velocity is less than the shear wave velocity in the specimen

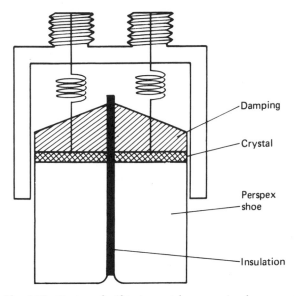

Fig. 4.22   Design of a 0° twin crystal compressional wave probe

restricts the choice when the specimen is aluminium or steel. For use on copper and cast iron specimens, Nylon wedges have been used.

In the Perspex wedge there will be reflected ultrasonic waves, and the wedge is shaped or damped so that these do not produce interference effects with the received pulse. Laminated Perspex has been proposed, and also specially-shaped shoes which provide a long transit path for the internally reflected energy. Transverse drilled holes in the shoe, filled with damping material, have also been used. Twin shear wave probes, with separate transmitter and receiver crystals, are also available.

It should be emphasised that ultrasonic probe manufacture is a highly commercial, competitive area, and this section can only enunciate the general principles of current designs. The most important points about shear wave probe design are that the beam angle of the shear waves is accurately known

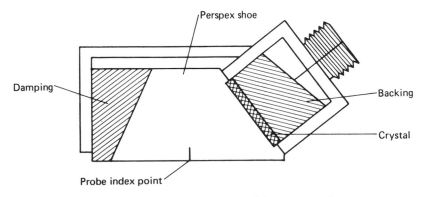

Fig. 4.23   Design of a single crystal shear wave probe

Table 4.2   Shear wave probe angles (degrees)

Nominal beam angle	Steel	Aluminium	Grey cast iron
35	35	33.0	23★
45	45	42.4	28
60	60	55.5	35
70	70	63.4	39
80	80	69.6	41

★In this case there will also be a compressional wave at 55°.

and the point of entry of the centreline of the beam into the specimen (the *probe index point* in Fig. 4.23) is also accurately known. These are two very important characteristics of any probe, which the ultrasonic equipment user needs to determine in the calibration procedures *for each individual probe*.

This design of angle probe, with a Perspex wedge of suitable angle, is also used for surface wave generation.

There are alternative methods of generating ultrasonic waves.

One possibility is to use the *magnetostriction effect*. Magnetostrictive materials change their form under the influence of a *magnetic field*, and the most useful magnetostrictive material in practice is nickel. A nickel rod placed in a coil carrying a current experiences a change in length, as a function of the current through the coil. A stack of plates of magnetostrictive material with a coil through them (Fig. 4.24) can produce an ultrasonic beam at right angles to the plate stack and the frequency depends on the thickness. Transducers of this type are useful for very low ultrasonic frequencies (<200 kHz), for testing concrete, etc., but have rarely been used for metal inspection.

Ultrasonic waves can be produced by *electromagnetic-acoustic (EMA) effects*, by direct electromagnetic coupling between an external coil and the electronic currents generated in the surface layer of the specimen. These EMA probes (EMATs) are becoming increasingly important in ultrasonic flaw detection. They will be discussed in more detail later (Section 4.3.2).

Ultrasound can be generated by lasers (Section 4.3.5), by the interaction of

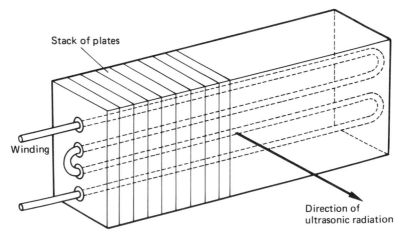

Stack of plates

Winding

Direction of ultrasonic radiation

Fig. 4.24   Magnetostrictive transducer

the optical wavetrain with the material, and the method has some inherent advantages over the conventional techniques, but as a very high proportion of ultrasonic flaw detection is performed with the types of probe shown in Figs 4.21–4.23, it is desirable to consider first the electronic circuitry used with such probes.

4.3.1   Pulse echo system

Apart from some older types of ultrasonic thickness gauge which used continuous waves with a resonance technique, and a few special techniques which measure the transmitted intensity, all other industrial ultrasonic flaw detection methods use the pulse echo system. In this method, which was first proposed by Firestone in 1940, and Sokolov in 1941, and demonstrated by Firestone (1945) and Sproule *et al.* (1945), the principle is as follows.

An electrical pulse is applied to the transmitter probe, which produces a short ultrasonic pulse which is propagated into the specimen through a couplant layer (Fig. 4.25). The same pulse triggers a timebase generator, so that the pulse of ultrasound starts to move through the specimen at the same time as a spot starts to move across the cathode ray tube, CRT, display screen.

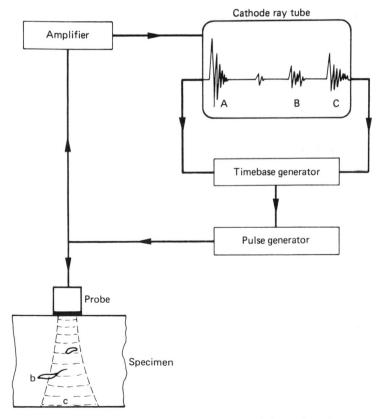

Fig. 4.25   Principle of operation of conventional ultrasonic equipment

Variations in voltage at the transducer due to the ultrasound wave are passed to the amplifier and applied to the Y-axis of the CRT to produce a transmission signal (A), which represents the shape of the generated ultrasonic pulse. The spot continues to move across the screen of the CRT as the sound pulse travels through the specimen until the ultrasonic pulse reaches a reflecting or scattering surface (b). The reflected portion of the ultrasound returns to the transducer, which vibrates, causing a small alternating voltage which is fed to the amplifier, where after amplification it is fed to the Y-plates of the CRT and produces signal (B), the echo pulse from the flaw. Further ultrasonic energy in the transmitted pulse may continue to the bottom surface of the specimen plate and be reflected back to the transducer, producing indication (C) on the CRT display, the bottom surface echo.

If the specimen is 100 mm thick, the travel-distance of the ultrasonic pulse would be 200 mm, which in steel will take only 33 μs: obviously, the display would be present on the CRT screen for such a short time that it would not be seen. To get an apparently steady display, the process needs to be repeated many times per second—typically, a pulse-repetition frequency, PRF, of 500–2000 pulses per second (pps) is used. At a PRF of 1000 pps, the time-gap between pulses is about 1000 μs, so that each pulse has ample time to reach a distant reflector and to be reflected back to the transducer, before the next pulse is emitted. At the end of each sweep, the CRT spot flies back to the left-hand edge and waits for the next pulse. This timebase scale across the CRT screen is adjusted to correspond with the thickness of the specimen being examined. If the specimen is much thicker, say 1000 mm, then the spot would travel across the CRT screen much more slowly, making a brighter trace and a PRF as high as 1000 pps would not be necessary. In fact, if too high a PRF was used, the subsequent pulse would be generated before the first returning pulse from the backwall of the specimen has returned, which would produce a very confusing display. Therefore a range of PRF-values is needed, and usually this is changed automatically as the depth control on the equipment is set. Modern flaw detectors can display the specimen thickness as a full-scale width on the X-axis of the CRT from about 10 mm steel to 5 m steel, by varying the spot sweep-time from about 4 to 2000 μs.

For probes in which the piezoelectric element is not in contact with the specimen surface, such as the twin crystal probe shown in Fig. 4.22, or immersion probes, it is convenient to display the ultrasonic pulses starting at the specimen surface rather than at the crystal, and a delay control on the timebase generator is used for this, starting the timebase after a preset delay. This same delay control allows the operator to look at part of a thick specimen on an expanded scale on the display. For example, any 20 mm out of a total specimen thickness of 200 mm can be displayed as full-scale width.

Since the ultrasonic signals may be weak or strong, the amplifier needs to have a calibrated gain control. For testing metals, ultrasonic probe frequencies from about 1 to 10 MHz are used, and much equipment is designed to use a slightly wider range of frequencies, from 250 kHz to 20 MHz. The gain control should be calibrated in decibels so that any signal can be increased or decreased in display amplitude by a known amount.

The amplifier may be broadband to cover a reasonable range of probe frequencies (for example 2–10 MHz), but switched frequency amplifiers are also used.

Circuits are usually incorporated to 'gate' any chosen part of the display and extract the signal within this gate for recording or to trigger an alarm or monitor circuit. The gate width and position, and the height of signal which is registered, are usually adjustable and are shown on the display screen.

Smoothing may be added to the display to remove random noise (sometimes called 'grass'). This can produce a more easily readable display, but it also removes some information from the display, and may reduce the resolution.

There are several possible types of display on the CRT screen (Fig. 4.26).

The type already described, with a linear timebase on the *X*-axis and ultrasonic signal amplitude on the *Y*-axis, is the most widely used, and is

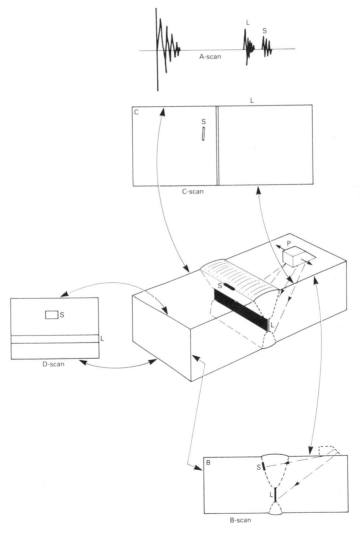

Fig. 4.26 A, B, C, D scans. A shear–wave probe (P) is assumed to be scanned transversely across the full width of the weld. The weld contains continuous lack of root penetration (L) and a slag inclusion (S)

known as *A-scan*. The ultrasonic signal can be shown 'raw', or rectified, or modified to show only a single maximum in each echo pulse. It is sometimes useful to examine the true pulse shape from the probe or in the reflected signal, and then the unrectified display is used.

There are alternative methods of presentation when a probe is scanned. In B–scan the apparent size and position of defects is shown on a cross-sectional plane normal to the surface and containing the beam axes of the probe during a single line scan.

In *C-scan*, a two-dimensional display of the test surface in plan view is shown. To produce this, the probe must be scanned mechanically over the surface in a regular raster and the defects are shown as bright patches in their correct plan positions. C–scan is widely used with immersion techniques, where the specimen is in a water tank and the ultrasonic probe is scanned parallel to the specimen surface. It produces a radiograph-like image, but provides no information on the through-thickness position of the flaw.

D-scan is a two-dimensional display of a cross-section of the specimen, along the line of scan of an ultrasonic probe, taken normal to the test surface and the beam plane, and therefore normal to a B-scan display. The apparent size and position of reflectors through the specimen thickness is therefore displayed.

P-scan is a commercial name given to a form of equipment which produces a simultaneous display of B and C scans. P-scan equipment is further described in Section 4.6.

In some equipment, the ultrasonic information is collected, digitised and presented as a computer read-out, or a graphical display. In addition to these forms of display, the digitisation of ultrasonic signals enables a form of tomographic image to be calculated by means of a computer program. (See also X–ray tomography, Chapter 3).

With some ultrasonic probes, particularly shear wave probes, part of the timebase can be obscured by transmission noise generated by reflection inside the probe, etc., and causes a dead space along the display trace at the start of the display corresponding to near-surface defects. This dead space can be reduced to a mimimum by probe damping, or, when using separate transmitter and receiver probes, by isolating the amplifier from the pulse generator and transmitter crystal, so that the initial phase does not appear on the trace.

It is emphasised that, as with ultrasonic probes, ultrasonic flaw detectors are commercial equipment which are continually being refined in specification, with the addition of extra features for the convenience of the user. The facilities described above are basic and many additional refinements are available in practice. The details of practical equipment characteristics and equipment calibration are discussed in Section 4.5.

There are a number of special designs of ultrasonic probe, all depending on the piezoelectric element as the source of ultrasonic energy, such as immersion and focussed probes, which are also used with conventional electronic, ultrasonic circuitry, which will be described in Section 4.6.

4.3.2   Electromagnetic–acoustic probes (EMATs)

Electromagnetic–acoustic phenomena can be used to generate ultrasonic waves directly into the surface of a specimen without the need for an external vibrating transducer and coupling. Similar probes can also be used for detection, so that a complete non-contact transducer can be constructed. The method is therefore particularly suitable for use on high-temperature specimens, on rough surfaces, and on moving specimens.

The EMAT principle is illustrated in Fig. 4.27 and 4.28. A flat 'pancake' coil of wire carries a radiofrequency current and a magnet (permanent or electro-) produces a steady magnetic field. The radiofrequency current induces eddy currents in the surface of a metal specimen and their interaction with the magnetic field results in Lorentz forces which cause the specimen surface to vibrate in sympathy with the applied radiofrequency current. Once ultrasonic vibrations have been generated in the specimen surface, they propagate into the specimen in the normal way and can be detected, on reflection from a flaw, by the same transducer, or a separate receiving transducer. For receiving the ultrasonic energy, the vibrating specimen can be regarded as a moving conductor in a magnetic field, generating an emf in the surface, so causing currents in the radiofrequency coil.

If the coil is positioned as shown in Fig. 4.27, with the magnetic field perpendicular to the specimen surface, the vibrations will be transverse to the direction of propagation—i.e. shear waves. If the magnetic field is parallel to the surface, as in Fig. 4.28, compressional waves are generated. The compressional waves are normal to the surface and angled beams cannot be generated with this arrangement. The clearance between the transducer and the metal surface affects the magnetic field strength; the strength of the eddy currents generated and the ultrasonic intensity fall off rapidly with increasing gap, but, working at 2 MHz, a gap of 1.0–1.5 mm has been found to be practicable provided that the gap is kept reasonably constant.

The received ultrasonic signal strength in EMA systems, compared with a conventional barium titanate probe, is down by 40–50 dB, but input powers can be increased and the signal-to-noise ratio is high. An EMA probe can also

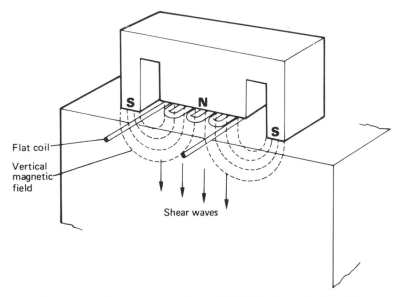

Fig. 4.27 Electromagnetic-acoustic transducer (EMAT) for the production of shear waves

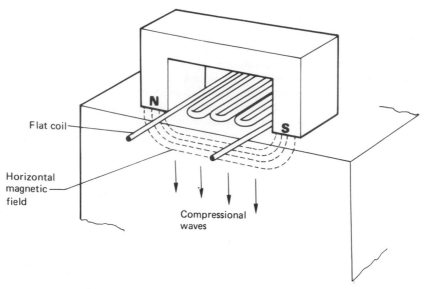

Fig. 4.28   EMAT for the production of compressional waves

be designed to generate surface waves, or Lamb waves. A zigzag radio-frequency grid of wires (Fig. 4.29) spaced half-a-wavelength apart is used, with the magnetic field tangential to the surface, so that the Lorentz forces are in the directions shown by the arrows. Thus two surface waves will propagate in opposite directions, outwards from the wire grid.

The design of EMA surface wave probes can be further refined. A

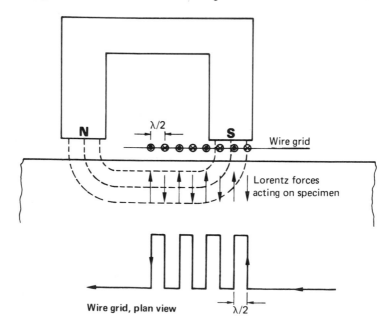

Fig. 4.29   EMAT for the production of Rayleigh waves

current-carrying coil or wire will generate its own magnetic field, which will itself react with the eddy current distribution in the specimen surface to produce Lorentz forces: thus, with sufficiently high currents in the transmission probe coil, it is possible to dispense with the external magnetic field. The wavelength of the surface wave so produced will be half of that when a steady magnetic field is used. The efficiency of ultrasound generation without an external magnetic field is, however, poor, but might be sufficiently high to be useful with pulsed techniques.

4.3.3   Focussed probes

There are technical advantages in modifying the ultrasonic beam shape by focussing the beam to a narrower cross-section, and there are two basic methods of doing this. Usually, such focussed probes are used in immersion systems, where the ultrasonic beam passes from the probe into water and then into the specimen. It is possible to use a curved plate of piezoelectric material, usually a ceramic material such as barium titanate, ground to the appropriate curvature, but the more common design is to have a lens of appropriate material cemented to the output face of a flat crystal. The curved crystal design tends to suffer from shear waves in the crystal, causing parasitic waves, and, of course, for contact techniques must be curved to match specific specimen curvatures.

A typical design of a lens-type focussed probe was shown in Fig. 4.15 with an aluminium lens and an epoxy material between the lens and the liquid, and a simpler design is shown in Fig. 4.30. The underlying principle of the lens design is exactly the same as for an optical system, but because the wavelength is very different, certain aspects of design require special consideration. The resolution will, generally, be limited by the same criteria as in optics, the limit being set by the diameter of the Airy disc, i.e.

$$1.22\lambda F/D$$

where λ is the wavelength, F is the focal length, and D is the diameter of the lens.

Image aberrations will be more significant than in optics, but the most important factor is that D is only a few wavelengths in size. The material of the lens should have an acoustic impedance to match the material on either side, to minimise reflections, but obviously must have a different ultrasonic wave velocity to the piezoelectric material.

To date, acoustic lenses for use in water have been mostly made of Perspex or polystyrene. Perspex has one of the lowest values of acoustic impedance for readily-available solid material, while at the same time it provides a refractive index (the ratio of the ultrasonic velocities) to water of about 1.8.

Spherical aberration accounts for considerable image unsharpness and there are secondary effects due to mode conversions at the lens interfaces. There is greater ultrasonic absorption off-axis because the thickness of the epoxy lens increases off-axis (Fig. 4.30), and this has the effect of spreading the focal spot along the axis. The probe shown in Fig. 4.30 has a minimum beam diameter of $D/3$ (to -6 dB), using a 20 mm, 5 MHz crystal.

Attempts have been made to construct focussed ultrasonic probes using a

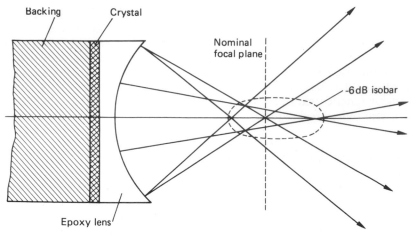

Fig. 4.30   Focussed beam probe with lens

Fresnel zone-plate directly in front of the piezoelectric element, sometimes called 'Axicon' probes. The advantage of a Fresnel zone-plate is that it has a short focal length and a high degree of focussing. A zone-plate consists of a series of annular rings, alternately opaque and transparent, with a central opaque zone. All the opaque rings have the same area, A_n, as the central zone

$$A_n = \pi F\lambda$$

and the zone radii, r_n, are given by

$$r_n = (nF\lambda)^{1/2}$$

where n is an integer.

   The aim is to produce a probe with a narrow field extending over a considerable distance. Alternative constructions are to use an annulus of piezoelectric material, or an annular electrode on a circular disc. Still another design, shown in Fig. 4.31, and attributed to Burckhardt, gave a focussed beam extending for 200 mm along the axis.

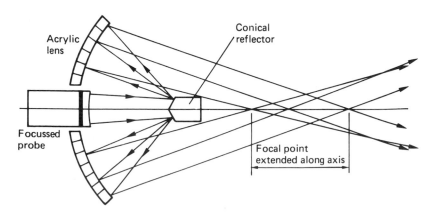

Fig. 4.31   Focussed beam probe

4.3.4   Focussed phased arrays

A still newer method of producing a focussed ultrasonic beam is the use of an array of transducer elements. If the phase- or time-delay to each element is controlled, the shape and direction of the ultrasonic beam can both be modified. The technique is well known in sonar and radar, and is widely used in medical ultrasonic testing.

A typical array might consist of one-hundred 10×0.3 mm silvered PZT strips cemented to a backing block side by side, of a thickness to oscillate at 4 MHz; an electronically controllable delay line is used for the adjustment of delays between elements. The basic principles are well documented and the principal problems seem to be due to constructional difficulties caused by variations in the effective acoustic characteristics of the individual piezoelectric elements on the backing material. Several arrangements are possible, as shown in Fig. 4.32, depending on whether a normal beam, an angled beam, or a focussed beam is needed; combined steering and focussing is also possible. However, the practical complications of mode conversions at the interface with the specimen affecting the interference patterns obtained by the phased excitation of the array, and the need to obtain quantitative rather than qualitative data, together with the general constructional problems, have so far prevented probe arrays being widely used in industrial applications.

Probes can be constructed either for vertical or for horizontal beam steering (Fig. 4.33), and by changing the element size and distribution, the beam directivity pattern can also be modified, so that the main lobe is optimised. For use with phased arrays, completely different electronic equipment is necessary, compared with the ultrasonic sets used with conventional probes.

Curved, multi-element arrays have also been built, which might effectively simulate the combined effects of several different angle probes. Each element produces a beam at a slightly different angle, but since the axis of each beam is directed radially, all the beams pass through the centre of curvature of the probe shoe. The probe elements are excited in groups, each group producing a concave wavefront, focussed at the emission point, but highly divergent. By using a precise time-sequence, an overall plane wavefront can be produced. With such a curved array, the side-lobes on the main beam, which are present with a steered (phased) linear array, can be minimised.

The advantage of the curved array is that it can be rapidly sequenced through a number of different beam angles, without the need for moving parts, and always has a fixed index point.

4.3.5   Laser generation of ultrasound

Ultrasound can be generated by laser pulses using the thermoelastic effect. The electromagnetic radiation is partly absorbed on the surface of the specimen, causing a sudden temperature rise, thermal expansion and a pulse of elastic waves. These are compressive waves parallel to the surface.

Typically a Q-switched Nd:YAG laser is used with power of about 5 mJ/pulse. With higher energies ablation effects permit other modes of ultrasonic wave to be produced (Le Brun & Pons, 1988).

Laser generation of ultrasound is contactless, and the short ultrasonic pulses

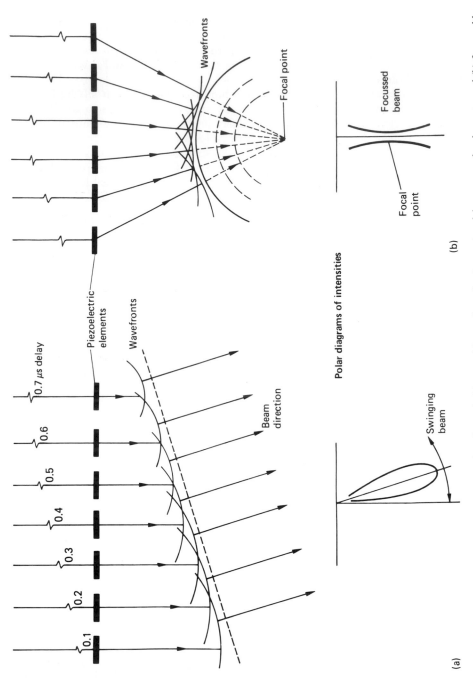

Fig. 4.32 Transducer with an array of elements (phased array): (a) steered beam, by using increasing delays across the elements, and (b) focussed beam

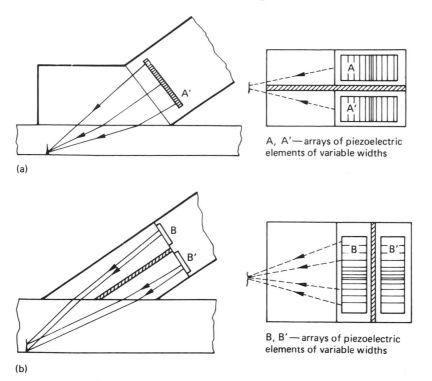

A, A'—arrays of piezoelectric elements of variable widths

(a)

B, B'—arrays of piezoelectric elements of variable widths

(b)

Fig. 4.33   Phased-array steered beam probes: (a) steering in vertical plane and (b) steering in horizontal plane

(c. 2 ns long) have been used for flaw detection in thin specimens. Laser generation has also been used for ultrasonic velocity measurements.

4.3.6   Probe variability

One of the major problems in ultrasonic NDT is the variability between probes of nominally the same design and construction. It has been claimed that this is due to inherent variability in the piezoelectric material, but the more likely cause seems to be the bonding of the piezoelectric layer on to the backing—i.e. the bond line. The width of the bonding layer and its uniformity in width have been shown to be critical in probe performance. Piezoelectric materials with a low acoustic impedance, such as lead metaniobate and quartz, are less critical in this respect. One solution to the problem might be to introduce a matching intermediate layer.

4.4   PROPERTIES OF ULTRASONIC WAVES

Most of the basic properties of ultrasonic waves which are important for industrial ultrasonic testing have already been described in Sections 4.1 and 4.2. These are:

(1)   reflection at interfaces,

(2)   refraction at interfaces,
(3)   mode conversion with angled incidence on an interface,
(4)   attenuation,
(5)   wavelength, frequency and velocity,
(6)   beam spread and the near and far fields.

It is convenient in the first instance to treat ultrasonic waves as having similar properties to visible light, and to consider an ultrasonic beam from a probe as analogous to a light beam from a pocket torch, but this analogy is strictly limited because in the ultrasonic beam, the beam is only a few wavelengths in diameter, whereas in a light beam, the diameter is many thousands of wavelengths. Secondly, mode conversion, as it is known with ultrasonic waves, is an effect which does not occur with light.

As already stated, a very large proportion of industrial ultrasonic testing employs piezoelectric contact probes, an A-scan display, and a pulse echo system with an ultrasonic frequency in the range of 2–5 MHz, and it is now necessary to re-examine the ultrasonic wave properties from the point of view of this type of application.

4.4.1   Probe output pulse

The shape of the pulse from an ultrasonic probe depends on many factors. The shape of the electrical pulse which shocks the piezoelectric element into vibration is a major factor which will be discussed in Section 4.5.1. The shape of the pulse also depends on the detail of the probe construction, as well as on the physical constants of the piezoelectric material. There is no such thing as a standard probe, or a standard probe output. A probe of any particular nominal frequency has a frequency spectrum, as shown in Fig. 4.2(c), and not a single frequency output. The shape of the output pulse (Fig. 4.2(a)) depends markedly on the backing to the piezoelectric element and the efficiency of adhesion of this backing to the crystal.

In consequence, the intensity distribution in the ultrasonic beam from a probe is also individual to a specific probe: for example, the magnitude of side-lobes on the probe output diagram can vary markedly between probes of the same nominal frequency, and if a contact probe is in uneven contact with the specimen, the angular characteristics of the beam may change.

The tendency at present is to use heavily-damped, broadband transducers having an output spectrum such as is shown in Fig. 4.34(b), in order to improve resolution. Newer transducer materials, such as lead metaniobate, have a higher internal attenuation and permit sharper pulses to be produced. It is therefore not strictly true to assign a single frequency to such probes, and calculated data such as DGS diagrams (see Section 4.6.2) needs to be modified. As a consequence of attenuation effects in some materials, the received pulse from a distant defect will also be changed in frequency content, compared with the input pulse, losing some of the higher-frequency components.

If the probe spectrum is not known, the operating frequency can be measured either by the frequency at which the peak occurs in the frequency spectrum of an unrectified pulse, or by counting all the complete cycles which lie inside the pulse length and dividing by the time-interval which they

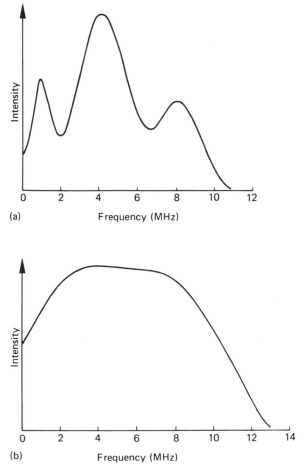

Fig. 4.34 Probe output spectra using a conventional pulse system: (a) 4 MHz resonance-type transducer and (b) 4 MHz broadband transducer

occupy. Usually all the pulses with an amplitude greater than one-tenth of the maximum are taken (Fig. 4.35). The resonant frequency of a loaded crystal, whether this loading is due to the backing, or the crystal-to-specimen contact, or both, is not the same as the resonant frequency of an unloaded crystal and cannot usually be determined by calculation: it must be determined experimentally. Frequency spectra can be plotted using a frequency spectrometer, but for ordinary ultrasonic flaw detection, the frequency spectrum itself is not used, although there have been attempts (see Section 4.7.1) to use the frequency spectrum from the reflected pulse, to characterise specific defect types (for example, different weld defects).

The practical probe frequency is quite often considerably different from the nominal frequency, due to inaccuracies in crystal thickness, lack of parallelism, and inhomogeneities in ceramic materials. The use of different electrical

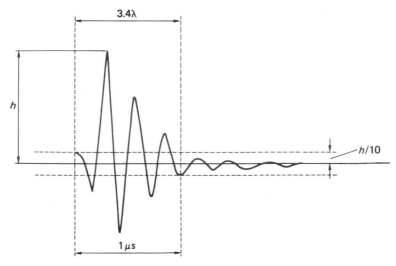

Fig. 4.35   Pulse from nominal 4 MHz probe (actual 3.4 MHz)

circuitry, such as the addition of tuning inductors, can also alter the pulse shape.

Several organisations have developed equipment to measure the complete sound field of a probe and some organisations, such as Harwell (UK), provide this data as a probe standardisation service. The measurements are obtained in one of two ways: either with an optical Schlieren interference method, or with mechanical scanning of the field in water using a small spherical reflector. The latter method produces quantitative data, but is comparatively slow. The mechanical bench must have extremely accurate movements and the reflector ball should be precisely spherical.

4.4.2   Calibration blocks

In the past, so many errors of interpretation were made in ultrasonic flaw detection due to variable probe and equipment performance that the need for better calibration procedures became very obvious. For a time, there was a large divergence of opinion between the value of the flat-bottomed drill-hole system, which had been developed by the USA aircraft industry, and a calibration block developed in Europe for ultrasonic weld testing. The flat-bottomed hole calibration system is mostly used for normal compressional wave probes, where the ultrasonic beam is positioned over the end-face of the drilled hole. It is largely a method of sensitivity setting, e.g. the sensitivity control is adjusted so that the $\frac{1}{16}''$ hole gives a full-scale echo.

It is not easy to produce accurately flat-bottomed, deep, narrow holes. Several modifications were proposed to the basic European calibration block, and today the design is standardised by the ISO, the IIW (which did much of the pioneer work), and almost every European standards organisation (BSI, DIN, AFNOR, etc.), as well as several other countries, such as Australia.

Several guides have been written (e.g. by the IIW, the Welding Institute, and the BSI) giving detailed procedures on how to use these calibration blocks, and some documents, such as the UK Electricity Supply Industry (ESI) Standards 98-2 to 98-7, detail similar procedures as well as acceptance standards for equipment and probe performance for all ultrasonic testing within that industry. These calibration blocks are designed for use with contact probes, not with immersion techniques.

The first calibration block was a $12'' \times 6'' \times \frac{1}{2}''$ block of steel with a $\frac{1}{16}''$ diameter hole through the $\frac{1}{2}''$ thickness at $3''$ from one end, but it has been replaced by the A2 block (sometimes called the IIW calibration block), which is now the most widely used. There is also an A3 block (also known as ISO block No. 2), which is a miniature version, designed for most of the same calibration procedures, but smaller and more easily carried: it is used mostly on site work. There are a number of other calibration blocks for special checks (e.g. IOW beam profile block, BSI A4, A5 and A7 blocks, and ESI blocks X and Y for tube testing applications), but the most important and widely used are the A2 calibration block (Fig. 4.36) and the A3 miniature block (Fig. 4.37). For precise details of dimensions and tolerances, the various standards should be consulted. Some users have added small refinements to the A2 block, such as a shallow quadrant slot on one side, which is not shown in Fig. 4.36.

Both the A2 and A3 test blocks can be made in various materials, so as to match as nearly as possible the material of the specimens to be tested. The standard steel block is in low- or medium-carbon ferritic (killed) steel, and the aluminium block in zinc-bearing fine-grain (DTD 5074) or low-copper, nickel-iron alloy (DTD 5084). Both blocks are tested for attenuation after machining and heat treatment.

The reader is referred to the calibration procedure documents referred to above for fuller details, but some of the more important uses of the A2 calibration block can be exampled.

(1) *Shear wave probe: probe index*   See Fig. 4.38. The probe is moved along the surface at A, until the maximum signal amplitude is obtained from the quadrant surface. The *true probe index* (the centre point of the ultrasonic beam from the probe) is then at the engraved mark on the test block, which is the

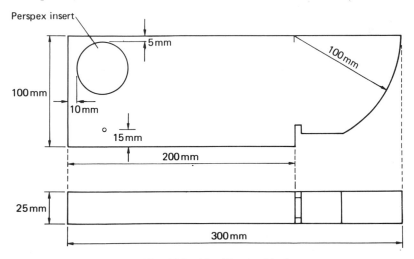

Fig. 4.36   A2 calibration block

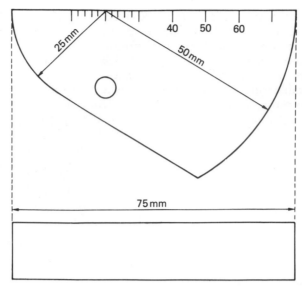

Fig. 4.37   A3 miniature calibration block

centre of curvature of the quadrant. This probe index point should be permanently marked on the probe, and should be accurate to within ±1 mm.

(2)   *Shear wave probe: probe angle*   See Fig. 4.38. Having previously determined the correct probe index point, the ultrasound beam is directed at the 50 mm hole with the plastic insert (position B). The probe is moved backwards and forwards until maximum echo height is obtained and then the angle of refraction β is read off. With the A2 test block, the probe angle should be determined to within ±1.5°.

(3)   *Depth resolution*   The original purpose of the slots in the A2 block was to provide resolution data. If a normal compressional wave probe is positioned over the 2 mm deep slot, the bottom of the slot is at 85 mm and the adjacent flat surfaces at 91 and 100 mm. These three distances should be easily resolved, the usual criterion being that the echoes are separated at half-maximum echo height or lower. For measuring higher resolutions which are attainable at high frequencies (4–6 MHz), special test blocks have been proposed.

(4)   *Shear wave probe: beam profile*   The beam profile of shear wave probes is

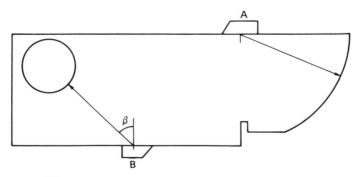

Fig. 4.38   Determination of probe index and probe angle

best checked with the A5 test block, as shown in Fig. 4.39, although a one-range measurement can be made with the A2 test block. The beam profile for a shear wave probe is found by using the probe on the 50 × 305 mm face for the horizontal profile and maximising the echoes from the tips of the target holes, then traversing the probe laterally to find the −20 dB drop position. The beam profile in the vertical plane can be similarly found, using the probe on the other surfaces of the block (Fig. 4.40).

It might be thought that the nominal angle of divergence of the beam, calculated from $\sin(\theta/2) = 1.08\lambda/D$ (that is, 20 dB reduction from the centre-line to the beam edge) was sufficient, but much more accuracy than this has been found to be necessary for critical flaw assessment in welds and the radiation pattern of a probe should be determined at several positions along the beam axis. Usually four points are determined just inside the periphery of the beam envelope, on the −20 dB level, anywhere along the beam, and these can be taken as lying on the surface of a cone which represents the ultrasonic beam.

Several other equipment and probe characteristics can, and must, be measured or checked before equipment is put into routine use.

(1) *Beam squint* This is defined as the angle between the side edge of a probe and the projection of the sound beam axis on to the plane of the probe face.

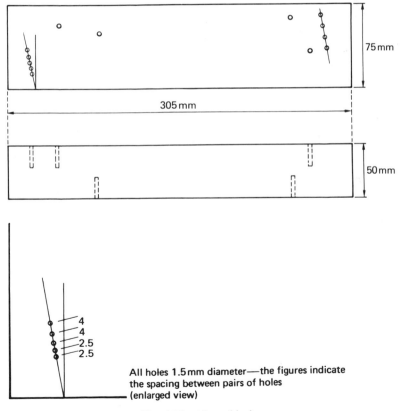

All holes 1.5 mm diameter—the figures indicate the spacing between pairs of holes (enlarged view)

Fig. 4.39 A5 test block

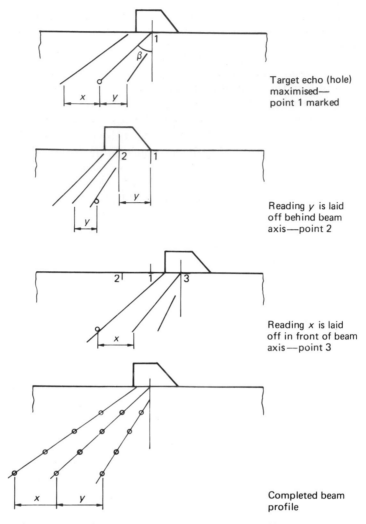

Target echo (hole)
maximised—
point 1 marked

Reading y is laid
off behind beam
axis—point 2

Reading x is laid
off in front of beam
axis—point 3

Completed beam
profile

Fig. 4.40   Determination of beam profile

Ideally, of course, it should be zero, and certainly not more than 1.5–2.0° on an acceptable probe.

(2)   *Signal-to-noise ratio* (electronic)   This should be at least 70 dB weaker than the echo from the 100 mm radius of the A2 block.

(3)   *Timebase linearity*   This can be checked by obtaining a series of echoes from a face, side or end of the test block, depending on the range to be used. The timebase should be calibrated to an accuracy of better than ±2% of the total range and many manufacturers claim ±1% for their equipment.

(4)   *Sensitivity*   The probe is placed to give a maximum signal from the transverse small hole in the test block and the calibrated gain control is set to give a low but clearly readable signal (e.g. 20% of full-screen height). The gain is then increased until the overall system noise reaches the same signal height, at the same range as the target hole. The first gain measurement provides a

check on the sensitivity of the ultrasonic equipment, and the difference between the two readings gives the signal-to-noise ratio. It is usual for application procedures to specify a sensitivity setting in terms of the echo pulse height on the screen from a drilled hole. For example, in some weld testing, a pulse 80% of full-scale height from a 3 mm side-drilled hole at the correct working distance is specified. The use of the side of the hole eliminates many of the problems in machining accuracy.

(5) *Measurement of dead zone* The dead zone is the region in the specimen, close to the surface on which the probe is placed, from which no direct echoes can be detected. The probe is placed on the 25 mm surface closest to the inset plastic block at either X or Y (Fig. 4.41). This check is only possible with compressional wave probes and is more of a check than a measurement.

Similar procedures for equipment and probe calibration with the A3 miniature block are set out in the various procedural handbooks and standards. For use on curved surfaces, slightly modified procedures have to be adopted, preferably on reference blocks made to the same curvature as the specimen to be tested. The geometry of the ultrasonic beam direction can become quite complex and it is recommended that an enlarged cross-sectional diagram of the set-up be produced.

Earlier American practice, which originated from the aircraft industry, was to use blocks with flat-bottomed holes, $\frac{3}{64}''$, $\frac{5}{64}''$ and $\frac{8}{64}''$ in diameter, drilled to different depths to cover the distance at which defects are to be detected in the specific application. A full set comprises 66 blocks, but at shop-floor level, probably only a few blocks are necessary at any one time. These blocks are for use with compressional wave probes.

More recently, the ASME reference block (Fig. 4.42) has become more widely used. It is made of the same thickness and grade of material as the specimen to be tested, and contains a single side-drilled hole, normally $\frac{1}{8}''$ in diameter. It is used to specify sensitivity settings and to produce distance–amplitude correction (DAC) curves so that an acceptable/rejectable signal-height level can be specified to cover a range of flaw depths. The shear wave probe is placed on the large, flat surfaces of the block, moved through positions 1, 2, 3 and 4, and the signal amplitude measured. The DAC curve can then be plotted.

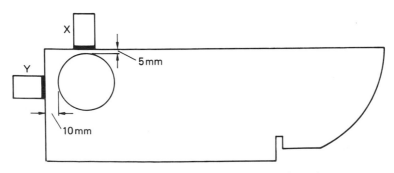

Fig. 4.41 Estimation of length of dead zone of a probe

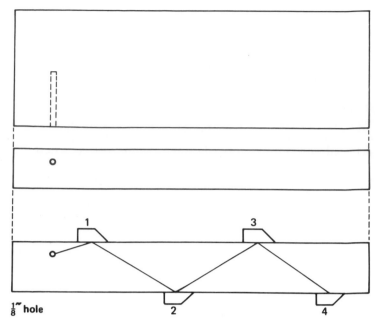

Fig. 4.42   ASME reference block

Distance–amplitude correction (DAC) curves

Apart from accurate data on the probe and equipment characteristics determined with the various types of calibration blocks and the specification sensitivity settings, there is a considerable amount of other data which should be obtained. Simple theory states that for a flaw of a given size and shape in the far zone of the probe, neglecting absorption effects, the echo height from the flaw varies inversely as the square of the distance of the flaw from the probe. It is possible to provide an amplifier circuit which can correct for the flaw distance effect, but in view of the complications due to absorption and scatter, it is more usual to use an accurately linear amplifier and provide a DAC curve; sometimes this DAC curve is superimposed on the cathode-ray tube screen. With the advent of computer-controlled ultrasonic equipment, distance–amplitude corrections can be applied automatically with a suitable program.

A suitable type of test block for preparing a DAC curve is shown in Fig. 4.43, the procedure being to adjust the probe position for a maximum echo from each hole in turn, having set the gain to obtain (say) 80% of full-scale height from the first hole. For short distances, the probe can be used on surface A. The sensitivity setting at which the DAC curve was obtained must be recorded. Figure 4.43 is for a shear wave probe, but an exactly similar curve can be obtained with the same test block for a compressional wave probe. In strongly attenuating material, it may be necessary to add a correction factor for attenuation.

With both flat-bottomed holes and side-drilled holes, it is not easy to obtain constant results, particularly with broad-band probes, due to the non-constant relationship between wavelength and target size.

A machined cone with a cone half-angle of 60°, which is a diffractor rather than a reflector, has been proposed as a standard reference (De Vadder, 1989).

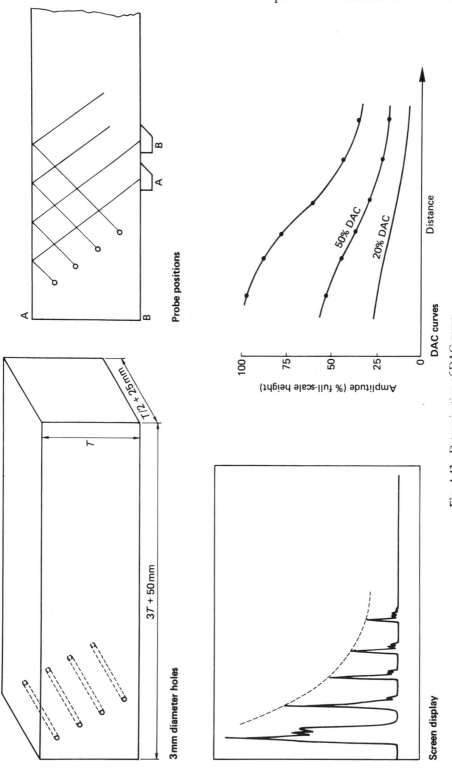

Fig. 4.43 Determination of DAC curve

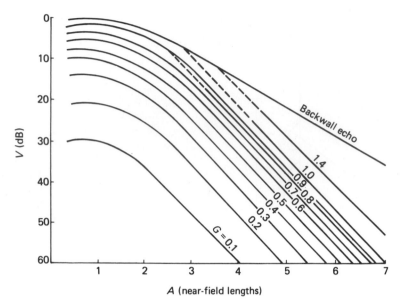

Fig. 4.44 Generalised DGS diagram, where A is the beam path in near-field lengths, G is the ratio of defect size/probe diameter, and V is the amplification required to obtain a defect echo amplitude equal to that from an infinite reflector

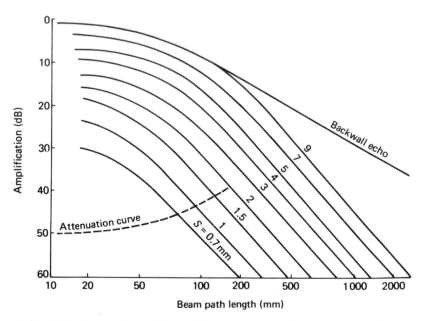

Fig. 4.45 DGS diagram for a 60°, 4 MHz shear wave probe, where S is the equivalent flat-bottomed hole diameter in mm

Distance–gain–size (DGS) diagrams (also sometimes called AVG diagrams)

These were first formalised by J Krautkrämer in 1958 and consist of plots of amplitude in decibels from a series of disc-shaped reflectors, with increasing distance of the probe, obtained in water. They are derived theoretically, and if plotted with distances in terms of the near-zone length, can be of general form, applicable to all probes. The loss due to water attenuation is allowed for, so the curves show data for any material, assuming no attenuation. DGS diagrams can also be derived for particular probes. Figure 4.44 shows the generalised curves for normal (0°) probes, and Fig. 4.45 shows specific curves for a 4 MHz angle probe with an 8×9 mm crystal size. DGS curves are used for defect sizing and will be discussed in more detail in Section 4.6.1 on weld testing.

DGS diagrams are used to estimate the *minimum size of a detected flaw*. They are calculated for a theoretically perfect reflector, which a natural flaw is not; they determine the size of the disc-shaped reflector which gives the same return echo pulse as the flaw. The less-than-perfect reflector—the flaw—will always be larger than the DGS diagram value.

There are a number of methods which are used to determine flaw size, besides the DGS system, and these will be discussed in Section 4.6.1. It will be obvious from the earlier part of this section that the competent user of ultrasonic flaw detection equipment needs to amass a lot of practical data about the equipment, particularly if attempting to size as well as detect flaws. It is not sufficient to accept the nominal parameters of probes and electronic circuitry, and no conventional ultrasonic equipment, as yet, automatically measures the sizes of flaws from the reflected ultrasonic energy. Both adequate training and considerable technical skill are essential for successful ultrasonic testing.

4.5   METHODS AND INSTRUMENTS FOR ULTRASONIC MATERIALS TESTING

Almost all ultrasonic non-destructive testing uses the pulse echo technique: applications of continuous wave methods are relatively rare. Practically all the earlier part of this chapter has been concerned with the ultrasonic probe and little has yet been said about the electronic circuitry to drive the probe and to collect and display the received signal.

Basically, the piezoelectric element in the probe is stimulated by shock excitation from a pulse produced in a conventional pulse generator. Pulses of about 500–1000 volts are needed, and to obtain a bright, persistent display, pulse-repetition frequencies of between 500 and 2000 pps are used. The piezoelectric crystal can be regarded as a capacitor and is shock excited by discharging a second capacitor into it, using a thyristor or avalanche transistor as the switching element to discharge the capacitor. The shape of the leading edge of the pulse depends on the turn-on characteristics of the thyristor, and the recovery time of the pulse. A second thyristor recharges the capacitor. Thyristors can switch very large currents and are fast enough for use up to and around 20 MHz.

The pulse generator is coupled to a timebase generator and to a conventional cathode-ray oscilloscope (CRO) as already described (Fig. 4.25). As the ultrasonic vibrations generated in the probe begin to propagate through the specimen, the electron beam in the CRO begins to sweep across the screen. The ultrasonic vibrations which are reflected or scattered in the specimen return to the probe (assuming single probe operation), and are converted to a voltage pulse, suitably amplified, and displayed on the CRO screen at the appropriate time-point. Synchronisation circuits, attenuators and ramp generators are also obviously necessary.

Returning to the pulse-generating circuitry, a high power output coupled into a low-noise preamplifier is needed to optimise the signal-to-noise ratio. The transmitter drive pulse should be free from ringing, to prevent excessive amplifier overload. The preamplifier needs to have a very short recovery time from overload, to enable received signals close to the transmitter pulse to be received without distortion.

The length of the pulse generated by the probe is a critical factor in flaw resolution, and short pulses also make it easier to distinguish defect signals in the presence of background 'grass'. Some of the newer ultrasonic techniques which will be described later, such as time-of-flight diffraction (TOFD), depend on precise timing circuitry, with a precision of a few nanoseconds, so that the excitation of the transducer must be equally precise. Similarly, the use of digital averaging techniques to improve the signal-to-noise ratio and the use of multiplexed or phased arrays of probes require improved time- and delay-stability. The frequency spectrum of the ultrasonic pulse is also changed by alterations in the electrical excitation, pulse shape and length. All these considerations have led to the design of improved transducer driver circuitry.

One of the most important techniques is to use a pair of power MOSFET devices, instead of the thyristor circuitry. MOSFETs are high-voltage analogue devices capable of switching high currents in a few tens of nanoseconds. Pulses of 200 V, 50 ns wide, into an impedance of 50 Ω can be generated, and when switched off there is a very high impedance with a leakage current of a few microampères only. It is known that the pulse response of a piezoelectric ceramic transducer can be controlled by shaping the drive waveform applied, and with a MOSFET circuit the drive pulse width can be adjusted to achieve damping of the transducer response by arranging for the trailing edge of the pulse to cause the transducer to resonate 180° out of phase from the resonance from the leading edge of the pulse. The drive pulse widths equals the resonant period of the transducer. If the drive pulse width is half of this value, the pulse is optimised for maximum output, but not maximum resolution. The effect is greatest on transducers with little mechanical damping, but even heavily-damped transducers show some improvement in performance. MOSFETs are majority carrier devices, unlike high-power bipolar transistors, and do not suffer from charge storage effects; they are relatively new devices and are being developed to handle higher voltages.

4.5.1   Probe configurations

Returning now to the ultrasonic probe, there are a number of basic probe configurations which are applicable to a range of testing problems (Fig. 4.46).

(1)   The basic pulse echo system using a normal (0°) compressional wave combined transmitter and receiver probe.

(2)   A through-transmission technique with transmitter and receiver on

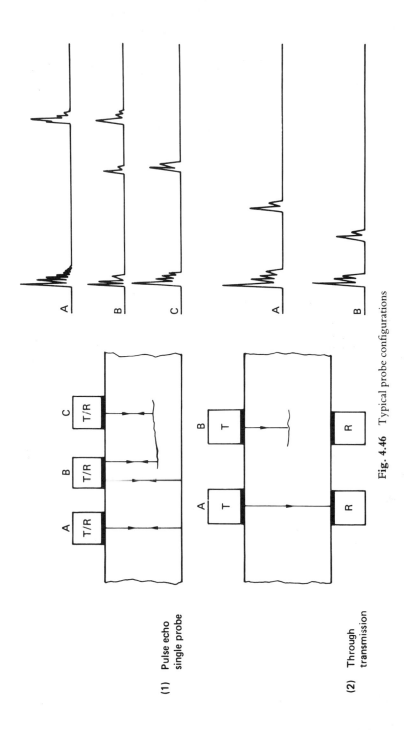

Fig. 4.46 Typical probe configurations

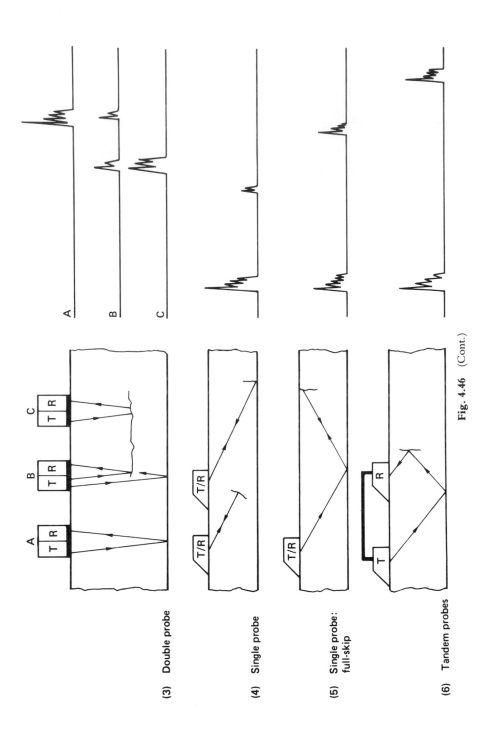

Fig. 4.46   (Cont.)

(3)   Double probe

(4)   Single probe

(5)   Single probe:
      full-skip

(6)   Tandem probes

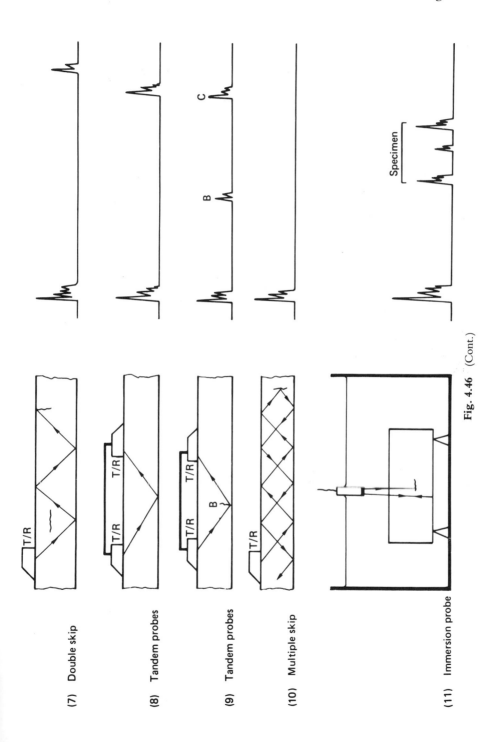

Fig. 4.46 (Cont.)

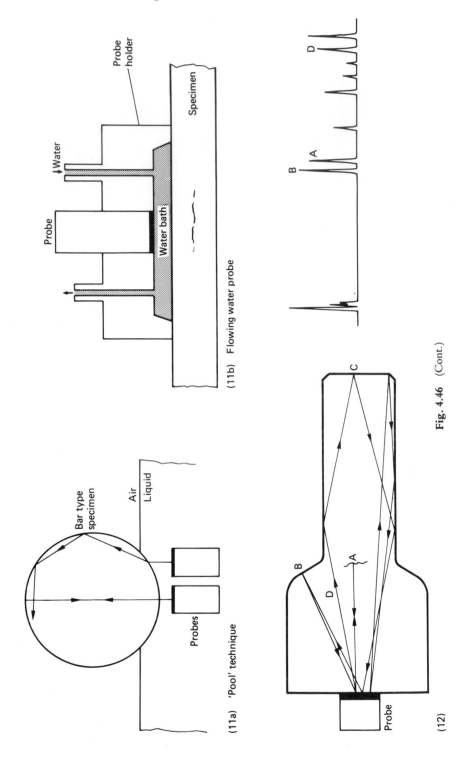

(11b)   Flowing water probe

(11a)   'Pool' technique

(12)

Fig. 4.46   (Cont.)

opposite sides of the specimen: if there is a flaw within the ultrasonic beam area (case B), the signal is lost or reduced.

(3)   A double-probe compressional wave system: normally there is no input signal shown on the display.

(4)   The basic shear wave pulse echo system.

(5)   The shear wave can be reflected off the lower surface—known as the *full-skip technique*. The *half-skip technique* is shown on the right-hand side of (4).

(6)   Separate transmitter and receiver technique to search for vertical defects: the two probes are mechanically linked.

(7)   The *double-skip technique* with shear waves. Note that the laminar defect would not be detected.

(8)   Another type of *tandem technique* with shear wave probes: both probes act as transmitters and receivers.

(9)   As for (8). The presence of a defect is indicated by an echo (B) and a loss of transmitted amplitude (C).

(10)   With a single transmitter and receiver probe, a vertically-oriented defect can easily be missed, due to the return echo missing the probe.

(11)   Water-immersion technique: the path length in water is large, so the distance between front and backwall echoes in the specimen may be rather small, unless 'beam expansion' is used on this part of the display.

(12)   With a large, complex-shaped specimen, there can be section changes which produce reflections, mode conversions and a mass of spurious indications on the trace. Most of the waves follow a longer path length than the direct reflection off the end C of the specimen, so many of the spurious indications are shown after the end-wall pulse, which makes them easier to identify.

All these, except (11), are contact techniques, used with a couplant between the probe face and the specimen. The probe may also have a protective membrane to minimise wear on a rough surface; compressional wave probes may have a delay shoe, as shown in Fig. 4.22, and this can be of a material such as Polyimide, which is heat resistant, for use on hot specimens. Soft-tipped transducers which require no couplant are available for use on composites and soft non-metallic surfaces. Usually, because of the interference losses, such probes are for transmission-mode operation only (Fig. 4.46(2)), and for low frequencies (0.5–1.0 MHz). Another method of overcoming the couplant problem is the roller probe (wheel probe). This consists of a soft plastic tyre completely filled with coupling fluid under pressure, with the piezoelectric crystal and its backing fixed to the shaft of the roller, so that it maintains its beam angle as the wheel is rolled over the specimen.

One of the problems encountered in ultrasonic testing is the detection and measurement of a crack which is approximately normal to the test surface. There is a range of possible probe orientations, as shown in Fig. 4.47, some of which depend on mode conversion on the back surface of the specimen. These are in addition to the rather cumbersome 'tandem' technique shown in Fig. 4.46, (6) and (9). CCC and CSS are double probe techniques, whilst SSCr and CCCr produce creeping waves on the crack surface. SCS and CCS are single probe techniques and it is possible to design a single probe transducer for a double mode (CCS) technique (Fig. 4.48). If ultrasonic testing can be done from the opposite surface to a surface-breaking crack, (Fig. 4.49), a high

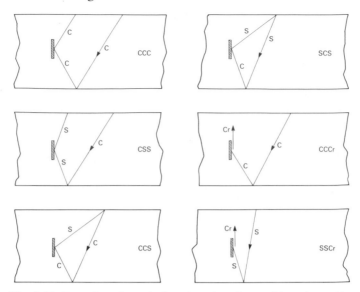

Fig. 4.47 Detection of a vertically-oriented crack. Possible probe positions.
C – compressional wave
S – shear wave
Cr – creeping wave

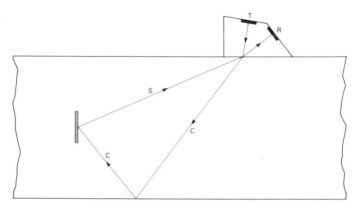

Fig. 4.48 Vertically-oriented crack. Double-crystal single probe technique

intensity normal longitudinal wave probe can be used to produce an echo from the tip of the crack, so that the height of the crack can be determined directly from the distance of this echo along the A-scan trace.

4.5.2   Coupling materials

With contact probes which are moved over the specimen surface by hand, a liquid is used to maintain ultrasonic contact between the probe (whether this is the piezoelectric crystal surface, a protective cap, or a delay block) and the specimen surface. All liquids have a much lower acoustic impedance than

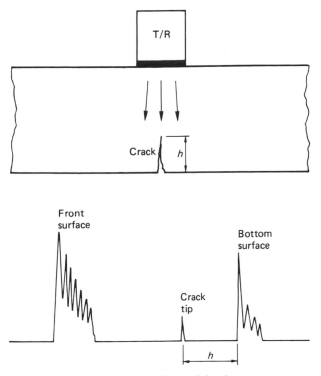

Fig. 4.49   End-on technique for crack height measurement

common metals so it is a case of choosing a liquid with as high an acoustic impedance as possible and suitable practical properties—e.g. non-corrosive, non-toxic, high viscosity, and inexpensive. Glycerine has the highest acoustic impedance of all common liquids, and is an excellent ultrasonic couplant, but is expensive. The commonest couplant liquid is medium-viscosity oil, the main disadvantage being its tendency to spread where it is not needed—on report sheets, clothes, etc. Silicon–oil can be used for hot surfaces, up to about 300°C. Thick oil, grease, or petroleum jelly can be used on vertical surfaces. A cleaner material, which is very successful, is a water-soluble paste, such as wall-paper paste (in the UK one proprietary name is 'Polycell'). If the couplant is electrically conducting as with water-based materials, it also acts as the front electrode with a quartz element probe.

All couplant layers reduce the sensitivity and broaden the width of the reflected pulses. In theory, if the couplant thickness is one-quarter of the wavelength (0.4 mm at 4 MHz) the multiple reflections in the couplant layer interfere, cancelling one another out, and cause a loss of energy back to the piezoelectric crystal when it is in the receiving mode. If the couplant thickness is half of the wavelength, there is no interference cancelling, but some additional pulse broadening due to the extra time-delay of the multiple reflected pulse in reaching the receiver. In practice, with a typical surface, the couplant layer does not have a constant thickness over the width of the probe element, and the main effect is to add another unquantifiable variable to the height of the reflected pulse echoes. If a piece of metal foil is placed in the couplant layer, with couplant liquid on each face, there is a considerable reduction in the multiple reflections and so in the signal broadening, but more 'gain' has to be used to maintain reflected signal height. Although in general liquids cannot support shear waves, it has recently been discovered that honey and some other commercial substances can be an effective couplant so that a probe generating shear waves directly can be used for scanning on a metal surface. Such a probe has advantages in the inspection of austenitic steel specimens, as the polarisation is not limited to one plane (Silk, 1988). With immersion probes and water-gap probes, the distance between the piezoelectric plate and specimen surface is so large that pulse broadening due to multiple reflections is eliminated, but such probes are not very convenient for manual testing.

There is a special problem with shear wave probes when these are used in weld testing. Often there is no backwall echo which can be used as a reference, so that the absence of any reflected pulse may mean either 'no flaws' or that no ultrasonic energy is getting into the specimen due to a lack of couplant liquid. It is possible to have a small, normal compressional wave crystal (A in Fig. 4.50) on top of an angle block which will only produce an echo from the bottom surface of the weld when couplant liquid is present. Effectively, this monitors the efficiency of the couplant.

For high-temperature ultrasonic testing, various proposals have been tried. A delay block (buffer) of heat-resistant material with a good ultrasonic transmission can be used between the piezoelectric element and the specimen: this can be liquid cooled. It needs to be in contact with the specimen for a few tenths of a second only—long enough to transmit and receive pulses. A couplant liquid is still necessary, and because the ultrasonic velocity and the attenuation in couplant liquids are markedly temperature sensitive, the beam angle of shear wave probes can be changed by temperature effects.

Buffer rod/transducer assemblies with momentary contact have been used on metals up to 2500 °C, but there is also interest in probes which can be immersed in couplant liquid at the operating temperature. Various liquid metals (low-melting-point alloys), such as eutectics of bismuth, zinc, lead and tin (Cerrobend) and non-tin eutectic (Cerrobase), have shown considerable promise in that the sound velocity does not change with temperature in the way that it does in an oil couplant, but there have been no reports of these liquid metals having been used above 250 °C.

The problem of ultrasonic energy losses in the couplant layer becomes of

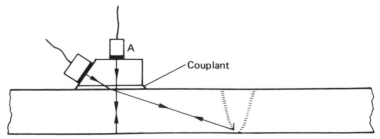

Fig. 4.50   Method of monitoring couplant efficiency when using a shear wave probe

major importance when ultrasonic attenuation in a specimen is being studied, and it is usual to work with extremely good surfaces to ensure that the couplant thickness is a very small fraction of the wavelength of the ultrasonic energy in the couplant. This is still not easy to achieve with high megahertz frequencies, where the ultrasonic wavelength may be only 100 μm (at 20 MHz).

4.5.3   Thickness gauging

One of the most important engineering uses of ultrasound is the measurement of metal thickness, from one side of a structure. Applications range from the simple checking of machined parts where it is not possible to apply a calliper or micrometer, to the measurement of pipelines in situ, to detect corrosion on the inner surface, with access from the outside only.

The early instruments for this purpose used the resonance principle. A probe was placed on the surface and a variable frequency oscillator was tuned to a thickness resonant frequency, as shown by an increase in the oscillator current. The amplifier resonance pattern was displayed on a CRT screen, which was calibrated in thickness and tended to be large so as to give a large scale length. A series of resonance peaks was shown and, by using harmonics, the measurable thickness range could be extended to about 100 mm. Such resonance instruments are now obsolete. The need today appears to be for small, portable, battery-operated instruments, measuring on a very small contact area (about 1.5 mm²) and measuring thicknesses from 1.0 to 300.0 mm to an accuracy of ±0.1 mm.

Modern thickness-measuring instruments all employ the time-of-travel measurement between the start pulse and the back echo, using pulse counting from a quartz clock. The instruments depend therefore on the ultrasonic velocity in the material under test being known, and the more advanced have a built-in microprocessor into which the velocity value can be keyed, either a known value or as measured on a calibration test piece. The display is direct-reading digital and, after calibration, is given as a true thickness reading.

With some digital gauges (Fig. 4.51), a twin crystal probe is used with an inclined beam, the idea being that the ultrasonic energy is focussed on to a very small area of the back surface and so can measure the depth to corrosion pitting, etc. The instrument automatically corrects for the slightly longer ultrasonic path length, and displays the correct wall thickness.

By using an EMA (electromagnetic-acoustic) transducer (see Section 4.3.2), a thickness gauge which requires no couplant, and which can be used on rough, scaled, or painted surfaces, is possible. The accuracy of measurement is not as great as with a digital thickness gauge, and the probe head is large, as it has to contain a magnet. Usually, EMA transducers are used with a conventional ultrasonic set with a CRT display, and an ancillary power supply for the magnet. Being non-contact, they are particularly suitable for use on hot pipework, and the off-stand of the probe from the metal surface can be up to about 2 mm. EMA probes have also been used for defect detection in hot steel products, but only compressional ultrasonic waves can be used at temperatures over 1100 °C. With EMA generation of shear waves, there is a marked temperature-dependence effect above 800 °C. The material to be measured must be electrically conducting, whereas digital thickness gauges can be used

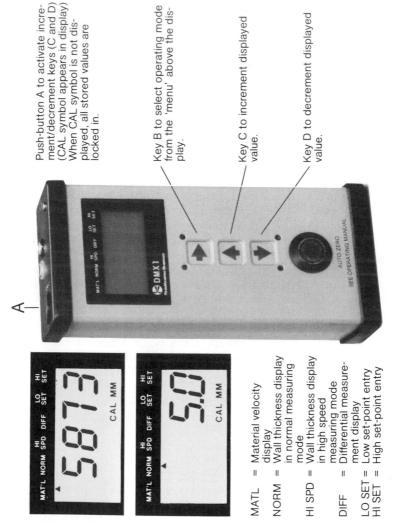

Push-button A to activate incre-
ment/decrement keys (C and D)
(CAL symbol appears in display)
When CAL symbol is not displayed, all stored values are
locked in.

Key B to select operating mode
from the 'menu' above the display.

Key C to increment displayed
value.

Key D to decrement displayed
value.

MATL   = Material velocity
            display
NORM = Wall thickness display
            in normal measuring
            mode
HI SPD = Wall thickness display
            in high speed
            measuring mode
DIFF   = Differential measure-
            ment display
LO SET = Low set-point entry
HI SET = High set-point entry

Fig. 4.51   Commercial portable ultrasonic thickness meter with built-in microprocessor

on any material which transmits ultrasonic energy and has a constant sound velocity value.

High frequency ultrasound can also be used to measure small thicknesses, e.g. down to 100 microns with 100 MHz.

4.5.4   Through-thickness methods

If a transmitter probe is placed on one side of a specimen and a receiver on the opposite side and these are scanned across the specimen, any variations in ultrasonic attenuation in the specimen will be registered. The method is sometimes called the 'obscuration technique'. Thus an internal flaw in the specimen will reduce the intensity of the transmitted ultrasonic energy as it comes between the probes.

In practice, the technique is difficult to apply, as the two probes must be maintained rigidly at a fixed distance and angle. Any probe movements relative to one another, or any variation in the specimen surface, will cause a change in the transmitted intensity. Consequently, applications of through-transmission techniques are comparatively rare. Material attenuation measurements can usually be made more consistently, and more accurately, working from one side of the specimen.

A few successful applications have been reported, using a pair of probes on a yoke. Usually these are placed to transmit an ultrasonic beam on to an inner surface, to be reflected back to the second probe (Fig. 4.52). Thus, if the pair of probes is traversed across a metal/rubber bond, or a brazed pipe joint, for example, any variation in the bonding interface shows up as a change in the transmitted ultrasonic energy. The disadvantage of the method is that it produces no information on defect location, but it is claimed that defects such as oxide layers in friction welds, which are not easily detected by conventional methods, can be found by through-transmission techniques. A pair of probes can be used on either side of a honeycomb or composite structure to monitor delaminations or disbonds. In one particular application, a 'dry scan' technique is used with two wheel-probes: these work at low frequency, without a coupling medium, and by using a continuous wave instead of a repeated pulse, enough energy is transmitted through the air-filled honeycomb to produce a detectable signal. Such techniques are particularly applicable to soft materials and absorbent surfaces.

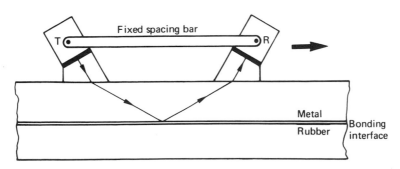

Fig. 4.52   Adhesive bond inspection

4.5.5   Automated and immersion systems

The basic ultrasonic flaw detection method uses a contact technique, with a hand–held probe and suitable couplant between the probe and the specimen. Variants of this technique are shown in Fig. 4.46(1) to (9), and the technique will be discussed in further detail in Section 4.6.1. The technique has obvious limitations such as coupling variations, unplanned probe position and angle changes, and recording problems (as well as some advantages of flexibility), and methods to overcome these limitations will be described under the general heading of automated ultrasonic systems.

With a specimen of predictable shape, such as a weld along a flat plate, or a circumferential weld in a large pipe, it is obviously possible to arrange for a probe to be moved mechanically over the necessary scan path to ensure coverage of the volume to be examined. The A–scan display from the probe can be gated to cover the relevant area in which the echoes from flaws might occur, and the signal from this area converted into an electrical or digital signal and recorded by means of a pen–trace on a chart, or stored as a number in a computer framestore. The timing circuit of the storage system must be synchronised to the mechanical movement of the probe. The probe scanning movements to examine a weld in thick material are quite complex, and it is more practical to use an array of probes, each feeding its signal into one pen of a multipen paper recorder. For example, one probe might be positioned to cover the root of a thick weld, one the middle region, and one the upper third of the weld metal. In order to detect planar defects such as through–thickness cracks, it has been found advantageous to have probes generating shear waves at different angles, for example 45°, 60°, or 70°, to have a probe positioned to detect transverse cracks, to have a pair of tandem probes, to have a separate compressional wave probe to monitor specimen thickness, etc.; equipment using as many of thirty probes has been developed for thick weld inspection.

One major practical problem is to ensure and maintain probe coupling. As stated earlier, with shear waves, no return signal can mean either no flaws or no couplant; it is therefore necessary to have a method of monitoring the coupling efficiency.

With compressional wave probes, one method is to dispense with the thin layer of oil or paste and deliberately space the probe front from the specimen surface with a small water–filled gap, through which water is continually flowing. This cannot of course be used with shear waves as these are not transmitted through a liquid such as water.

Several variants on this method are possible.

(1)   The specimen can be totally immersed in a water tank and the probe traversed over it, using a large water spacing—this is the 'immersion method'. The transmission pulse is then several centimetres of water ahead of the front surface pulse and can be gated out, with the space between the front and back surface echoes expanded to occupy the full-scale distance. The probe can be on a universal joint from a gantry traversing the immersion tank and can be programmed to follow the contours of a complex specimen such as a casting and maintain the ultrasonic beam at a fixed, chosen input angle and distance.

(2)   With a water column or jet between the probe face and the specimen surface, the water column transmits the ultrasonic beam, and the need for a full

immersion tank is obviated. Multiple, multijet systems have been built for both pulse echo and through-transmission testing. The water flow in the jet needs to be free from reflecting air bubbles and to be non–turbulent.
(3)   Bar and tube stock can be fed through a small immersion tank containing the probes, using water–tight end seals.
(4)   An ultrasonic probe with a water–jet can be held and moved by a computer–programmed robot.
(5)   A hand–held ultrasonic probe can be coupled to a plan-position indicator by a pair of potentiometer positioning arms which control the position of a pen on a chart recorder. Again, the probe movement parameters can be digitised and computer recorded.

All these methods of moving the ultrasonic probe mechanically depend on the accuracy of the mechanical systems, to produce reproducible results. The ultrasonic information can be collected from a gated A–scan trace, or a series of gates can be used, or it can be presented in B-, C- or D-scan. The tendency today is towards computerised equipment, using the computer to control the probe movement, to store the data after digitisation, and to process the data. A multiprobe automatic system produces an enormous amount of ultrasonic data: so much, that methods of handling it and presenting it in understandable form are still only in the early stages of development.

Many automated systems have been built for specific applications and the number of these is constantly increasing. Many have acronyms as commercial names, and the two large fields of application are weld inspection and aerospace component inspection. A few systems will be briefly described to show the type of present-day equipment, but it is not possible to include more than a small sample of what is available.

DRUID (*digital remote ultrasonic imaging device*) takes data from a single probe scan and generates B, C and D scan images. Amplitude information is coded into a number of levels and each band is represented by a different colour on the display. The data is recorded in digital form on a floppy disc and can also be processed off-line (Johnson, 1984).

P-SCAN   This is a weld inspection system, designed and built by the Danish Welding Institute. It can be a small manual scanner or an automated system. The probes, one on each side of the weld, are each moved over a preprogrammed scanning pattern. The echoes are measured with 1 dB resolution, without a preset detecting level and all echoes are recorded on magnetic tape. Projection images are then calculated and presented in simultaneous C- and B-scan formats on a display monitor, so that the 'standard' display is plan view (C-scan), side view (B-scan), and signal level in dB, together with weld and position identification numbers. A-scan display is also possible and all the displays can be at any chosen sensitivity level. The echoes obtained by scanning from the two sides of a weld (not the two sides of the welded plate) are distinguished on the display and need to be taken together for a complete interpretation of certain weld defects. All images are automatically recorded on tape or as hard copy.

AUGUR   This is a system developed for heavy plate testing by IzfP, Saarbrücken. It takes in data from 288 channels of ultrasonic pulse testing, namely maximum echo amplitudes, defect pulse transit-times, and backwall echo transit-times. Each probe assembly consists of one transmitter and three receivers and there are three units of 32 probes. The normal presentation is C-scan and the area of any defect can be calculated, to be used against an acceptance/rejection criterion according to various standards such as ASTM-A-435 or EURONORM-160. For example, ASTM-A-435 rejects any flaw that cannot be encompassed within a 75 mm diameter circle and AUGUR can assess the stored data against such a criterion. With the number of probes used, the data has to be compressed for processing, in order to operate at the design inspection speed of 1 m s^{-1} of weld length.

ZIPSCAN-3 developed originally by the NDT Centre, Harwell, is a digitally-based data storage and data analysis device which can be operated with any type of probe system—pulse echo, multi-probe, through-transmission, crack tip diffraction, tandem, etc. The r.f. waveforms are captured and can be displayed in real-time in A, B, C, D scans, quantised and if required, in false colour.

The TOFD and pulse-echo modes are complementary and together allow the data to be presented as an easily-understandable cross-sectional image of the specimen. A major application has been to detect and measure undercladding cracks in a structure with austenitic steel cladding on a thick ferritic base.

Using off-line processing, there are programmes for SAFT, spectral analysis, frequency averaging, DGS and dB drop sizing.

ALOK (*amplitude transit-time, locus curve*) consists of a family of equipment developed by IfZP, Saarbrücken, Germany, which similarly collects and processes data from either a series of probes, or a series of probe positions. The signals are not gated, so all the A-scan data from each probe position is utilised in digital form: noise can be eliminated by signal integration, transit-time (probe-to-flaw distance) locus curves are drawn, and then the computer is used to reconstruct an image of the defect, by triangulation, from a series of intersecting transit-time locus curves. A highly-damped type of probe with a broadband transducer is normally used.

ALOK has the possibility of using inspection systems with up to 32 different transducers, including different angles of incidence. A series of different reconstruction algorithms is available, and a two-dimensional image can be constructed. A, B, or C scan presentation is possible and the equipment operates off a 32-bit computer with both hard and floppy disc storage.

The purpose in detailing these examples of current automated ultrasonic testing is to emphasise the problems in handling multiprobe equipment: there is such an enormous quantity of ultrasonic data that if a rapid specimen scannning-time is to be achieved, data compression is essential. Effectively, a lot of ultrasonic information has to be ignored. The major lines of development are in the digitisation of data and computer programming to handle this data, and this must obviously be planned around each specific application. On the ultrasonic side, the problems are those of matching individual probe performances, monitoring probe coupling, standardisation of operating conditions, etc.

A major advantage of all digital recording systems is the permanent storage of data and the flexibility of computer programming for data analysis. The use of computers enables such features as built-in memories, digital or analogue images, multiple gating, and DAC correction to become standard features, and in many applications it is possible to have a cross-section of the specimens on the screen image, with the images of defects superimposed in their correct locations.

A wide range of signal processing is available for the improvement of digital data. Examples are:

using a matched filter for S/N enhancement
inverse filtering for resolution improvement.

Many of these come from other fields such as radar, sonar, etc.

4.5.6   Time-of-flight diffraction (TOFD) methods

The normal pulse echo method of ultrasonic testing uses the transit time of the pulse to measure the distance of the defect from the probe, and the amplitude of the reflected pulse to provide some information on the size of the reflecting defect.

There is an alternative, relatively new technique, which is entirely different in principle, which is now known as TOFD (time-of-flight diffraction). The geometrical theory of diffraction at an edge was originally developed for optics

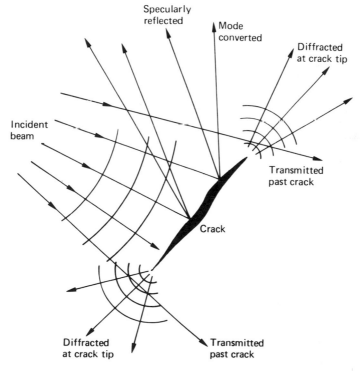

Fig. 4.53   Principle of time-of-flight diffraction (TOFD)

by Keller (1957) and first described and utilised for ultrasonic waves by Silk (1977).

   If one considers a broad compressional wave ultrasonic beam incident on a planar defect of finite length (Fig. 4.53), some of the ultrasonic energy is reflected as a compressional wave, some energy is mode converted to shear waves, some passes round the defect and does not interact, some is scattered from the defect face, but some is also diffracted at the tips of the defect. These diffracted waves arise at any major angular structure along the defect, but for inspection purposes it is those from crack tips which are of major interest. The mathematics of the formation of these diffracted waves is complex, because of the large number of ultrasonic modes of vibration which are possible and the multiple mode conversions and reconversions within the defect (Temple, 1988). For a crack as shown in Fig. 4.54, and a transmitter probe T, diffracted ultrasonic energy from the crack tips can be detected at R. The diffracted ultrasonic waves spread over a large angular range. The transmitter and receiver probes are located symmetrically from the crack tips, so that if the diffracted echo can be detected, a simple measurement of ultrasonic path length enables the depth of the crack to be calculated. If the two probes are $2S$ apart, the transit time t (μs) from the crack tip at depth d (mm) below the surface is

$$t = \frac{2(S^2 + d^2)^{1/2}}{c} + 2t_0$$

where c is the compressional wave velocity (in mm μs^{-1}), S is in mm, and t_0 is the time–delay in the probe shoe and couplant (in μs).

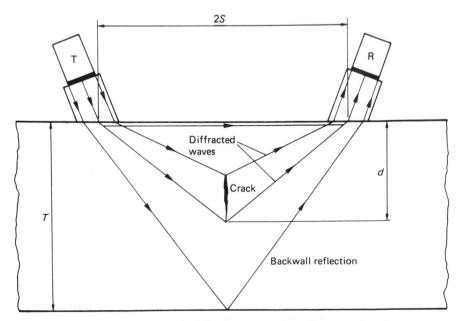

Fig. 4.54   TOFD method for the measurement of an internal crack

The probe time-delay t_0 can be determined from either the lateral (surface) wave or the backwall reflection (transit times t_1 and t_b):

$$2t_0 = t_1 - 2S/c$$

$$2t_0 = t_b - 2(S^2 + T^2)^{1/2}/c$$

where T is the specimen thickness.

The velocity can also be derived:

$$c = \frac{2(S^2 + d^2)^{1/2} - 2S}{t_b - t_1}$$

so that the depth of the defect is related to the measured time of arrival of the diffracted signal:

$$d = [(c/2)^2(t - 2t_0)^2 - S^2]^{1/2}$$

If the crack does not extend vertically, or is not surface breaking, the transducers have first to be moved to detect a signal and then, with the transducer spacing held constant, traversed across the crack to determine the minimum travel-time, to ensure symmetry over the crack tip. For a non-surface-breaking crack, the traverse of the probes has to be repeated with two different spacings.

It can be shown that the diffracted signal is greatest when the crack tip interacts with the centre of the ultrasonic beam from the transmitter probe; there is also a possibility that ultrasonic scattering from microflaws and strained metal immediately ahead of the crack tip might contribute to the received energy, but this contribution is thought to be very minor. The usual geometry for crack measurement is a low-angle beam and here diffraction is likely to be the dominant effect.

Most of the work done on this technique has been with compressional waves, using wedges to obtain the necessary input angles. Shear waves with a shallower refraction angle penetrate deeper into the material, so that the minimum detectable crack depth with a probe-spacing of 80 mm is about 70 mm, and shear waves have rarely been used for TOFD. A combined shear/compressional wave arrangement is possible and would give coverage over the range $d = 25$ mm to $d = 100$ mm, for a probe-spacing of 80 mm.

The technique proposed by Silk and colleagues records the appropriate parts of the A-scan display, after digitisation and averaging, and presents this data as a D-scan display (Fig. 4.55). The Y-axis has to be corrected to give defect depth. Synthetic aperture focussing (see Section 4.7.3) can be used to improve the accuracy with which the defect length along the scan direction can be measured, by utilising the information in the 'tails' of the signals. On artificial slits, accuracies in measured depths have been as good as ± 0.25 mm, and for cracks with tips more than 10 mm below the surface, the accuracy of measurement is ± 1 mm. It must be emphasised that whereas conventional ultrasonic pulse echo flaw detection relies on measuring signal amplitudes and is affected by flaw orientation, flaw roughness, etc., in TOFD, the accuracy is dependent only on the error in time measurement between two ultrasonic pulses, and the recognition of the leading edge of the signal. Transit-times may be affected by surface roughness and couplant variations, but can be measured to an accuracy of a few nanoseconds.

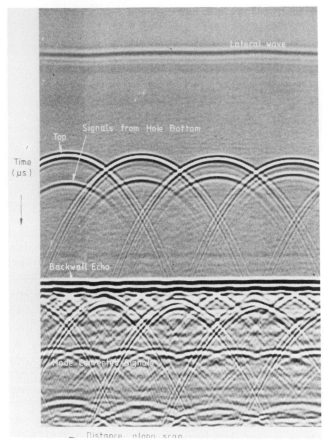

Fig. 4.55 Uncorrected D-scan from a test block with side–drilled holes of diameter 3, 4, 5 and 6 mm at a depth of 50 mm. Reproduced by permission from Carter P, *Brit J NDT*, **26**, 256, 1984

The practical problems in using TOFD are those of recognising the diffracted echoes in the presence of extraneous signals. In real specimens there is seldom one isolated single defect and echoes from other small volumetric defects can mask the relevant echoes. A useful pointer to the recognition of the diffracted echoes, as shown on an unrectified A–scan, is that the signal from the upper tip of a crack usually shows an 180° phase reversal, as does the backwall signal, whereas the diffracted signal from the lower tip, like the lateral (surface) wave, does not show this reversal.

If the pair of probes is not symmetrical over the crack, the tip of the crack will lie somewhere on an elliptical locus, which means that the location of the crack is subject to a considerable error, although the depth error will still be quite small. If one of the probes is made a combined transmitter and receiver, the crack location can be found accurately.

The main limitation to the TOFD technique still remains the identification of the diffracted echoes, particularly on a crack with a jagged surface, from which there may be secondary diffracted waves. Signal averaging is usually

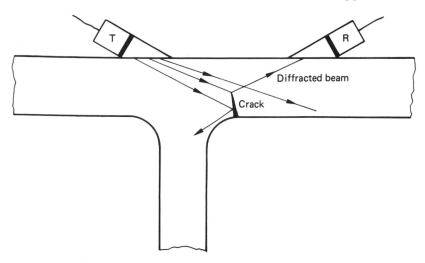

Fig. 4.56 TOFD applied to a T-junction weld

necessary for the relatively weak TOFD signals, particularly in a noisy environment.

With some crack geometries, for example on fillet welds in thick plate, it is possible and advantageous to apply TOFD techniques from the opposite side of the plate to the crack opening (Fig. 4.56). Variants on the basic two-probe TOFD technique, using one fixed probe and one traversing probe, or using skewed probes are possible and are advantageous for measuring irregular-shaped flaws.

Another method of using ultrasonic diffracted energy is to concentrate a focussed beam on to the crack tip and measure the amplitude variation as the angle between the axis of the beam and crack varies, the FET technique (Bonami & De Vadder, 1986).

4.6 APPLICATIONS

There is an enormous range of applications of ultrasonic testing, and in this section some of the more important applications will be outlined, particularly those that involve principles which have not yet been discussed in detail.

4.6.1 Weld testing

A major application of ultrasonic flaw detection is the inspection and quality control of welds. It is in this field that manual testing is most widely used and where defect-sizing is widely employed, and although many of the procedures to be described are also used in other applications, they will be described here in terms of the testing of butt-welds in ferritic steel, in the thickness range 6–200 mm. Austenitic steel is a special case, which will be discussed in Section 4.6.3.

Nearly all large industrial countries have produced national standards on ultrasonic weld testing (e.g. BS:3923, Parts 1, 2 and 3: *ASME Boiler and Pressure Vessel Code, Section XI*, and AS (Australia) 2207) and organisations such as the British Welding Institute (1971) and the International Institute of Welding (1977) have produced detailed procedural handbooks.

Most of these examinations are carried out with a hand-held shear wave probe and this must be calibrated, together with the equipment, using the calibration blocks and test procedures outlined in Section 4.2.2. At a minimum, the timebase linearity (±1%), the amplifier linearity (±1 dB), and the calibration gain control (±1 dB) on the equipment should be checked; the probe resolution, index point, frequency, shear wave beam angle, and the beam profile of each probe must also be known. The signal-to-noise characteristics of the probe-plus-equipment should be checked, and for some applications it is desirable to know the length of the near-field dead zone.

On the weld itself, it is essential to know the weld preparation (the shape of the joint) and to have a marked datum line showing the weld centreline. As much information as possible should be obtained on the type of material, heat-treatment, welding procedure, and surface condition. A cross-sectional drawing showing the fusion faces, the surface contours, and the actual thickness is an essential preliminary requirement, particularly on complex structures such as nozzles and cross-over pipework. The metal surface where the ultrasonic probes are to be applied must be free from scale and spatter and reasonably smooth: a surface finish not rougher than 4 μm(R_a) is desirable.

Usually, preliminary ultrasonic examination of the parent metal adjacent to the weld is made, using a compressional wave probe, to detect any laminations or lamellar tearing through which the ultrasonic beam might have to pass during testing of the weld metal. If the true position of the weld root bead is in doubt, such as in applications where the inside surface is inaccessible, this must be accurately located either by a compressional wave probe on the weld surface, or if the surface is rough, by a pair of shear wave probes on either side. The precise location of the weld bead and the centreline is vital to success in correct interpretation.

The most important part of a weld, from the point of view of significant defects, is the weld root and this is normally examined first. Lack of root penetration, lack of root fusion, and basal cracks can occur in this region, and must be distinguished from an undressed root bead. On single-sided welds, a separate scan should be carried out whenever possible with the probe positioned against a guide bar (Fig. 4.57(a)) at the exact half-skip distance to the root.

A 45°, 50°, or 70° shear wave high-resolution probe should be used, the angle depending on the weld thickness and the geometry restrictions. Accurate probe calibration and accurate probe positioning are necessary to separate the echoes from the edge of a root bead and a root crack (Fig. 4.58). The same basic principles apply to double-Vee welds, for detecting lack of root penetration (Fig. 4.57(b)). The main volume of the weld metal is examined by moving the probe between positions A and B (Fig. 4.59) as the probe is traversed along the length of the weld seam. Position A covers the lower portion of the weld at half-skip distance and position B the upper portion at full-skip distance. Additional traverses are necessary to detect transverse defects (Fig. 4.60).

Special slide-rules, which are effectively location diagrams on which an

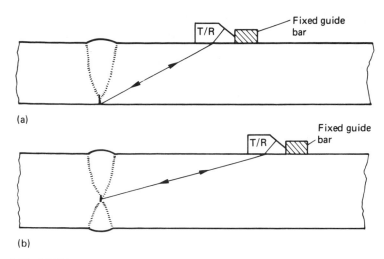

Fig. 4.57   Weld inspection: probe positions for (a) single-Vee weld and (b) double-Vee weld

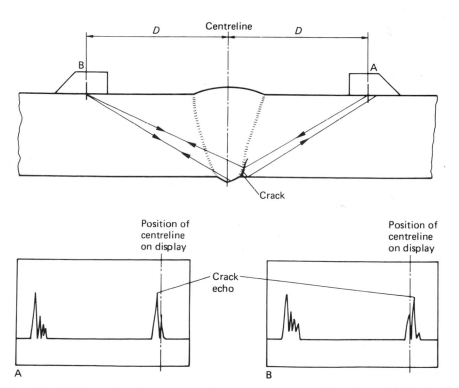

Fig. 4.58   Weld inspection: detection of crack or lack of fusion at the edge of the root bead. From side A, the flaw echo will generally show slightly greater amplitude than a root bead echo and will be slightly in front of the centreline position; from side B, the flaw echo is slightly behind the centreline position. The flaw echo will probably persist when the probe is moved in or out from the weld seam

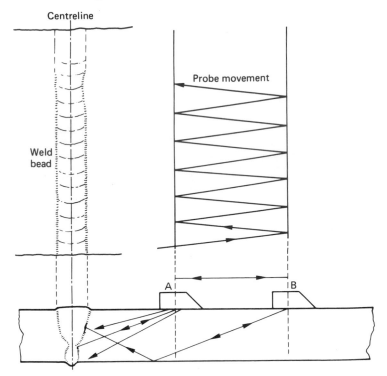

Fig. 4.59   Weld inspection: probe scanning pattern with manual inspection

enlarged image of the weld is positioned, are available, to assist in converting the ultrasonic measurements into flaw locations. For welds up to 100 mm thick, it is recommended that probe scanning is carried out from both sides of the weld on one surface. For thicker welds and critical applications, scanning should be carried out from both sides *and* both surfaces, whenever possible.

These are techniques for weld defect location and often the location of a defect automatically identifies the nature of the defect. More information of the nature of the defect can be obtained by observing the changes in the pulse shape as the probe is moved across the defect, or as the probe is rotated slightly. This 'dynamic image' is, to some extent, a function of the height of the defect, but also provides other information. Thus a planar defect such as a smooth crack or a lack of sidewall fusion will give a markedly directional indication, disappearing rapidly with change of beam angle, whereas the response from a similar-sized slag inclusion is more persistent, as well as being different in shape. It has been suggested that the pattern of the dynamic echo, as the probe is moved, can be grouped into a few standard forms (Fig. 4.61). Probe movement can be in two directions: thus, if pattern (b) is obtained with probe movement in two directions at right angles, the flaw is likely to be a large, planar flaw, approximately at right angles to the surface.

These general procedures for flaw location have been outlined only for the comparatively simple case of a flat plate butt-weld. For curved surfaces and structures such as nozzle welds, the principles are identical, but the practice is

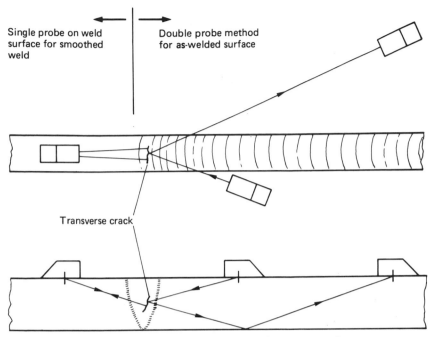

Single probe on weld surface for smoothed weld

Double probe method for as-welded surface

Transverse crack

Fig. 4.60   Weld inspection: detection of transverse cracks

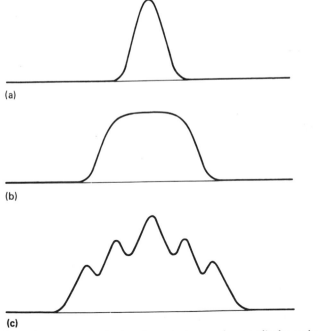

(a)

(b)

(c)

Fig. 4.61   Echodynamic patterns obtained as the probe is moved perpendicular to the weld seam: (a) due to a point defect such as a gas pore, (b) due to a planar defect such as a smooth crack face, and (c) due to a series of small pores or inclusions

more complex. Considerable procedural detail is included in the standards referred to earlier, such as BS:3923, and in documents produced by bodies such as the Electricity Supply Industry. Very similar procedures have been developed for flaw identification in forgings, based on a number of 'standard' echo patterns (Nussmüller, 1987).

For applications where access is difficult or impossible remotely-controlled devices such as pipeline 'intelligent pigs' or crawlers, carrying arrays of many ultrasonic probes have been built. Typical applications are corrosion measurement in a pipeline, or work in a radioactive environment. A 'pig' may contain several hundred sensors and a vast amount of data has to be stored and processed, necessitating considerable data compression. The relative advantages of ultrasonic and magnetic flux leakage methods depend on the wax/water/gas content of the oil in the pipeline.

4.6.2   Defect sizing

The sensitivity setting of ultrasonic equipment is often referenced to a DAC (distance–amplitude correction) curve, which can be obtained from a series of 3 mm side-drilled holes in a reference block (see Fig. 4.43). According to the examination standard required, the sensitivity is set at the level where 'grass' is just visible up to the maximum testing range, or the gain control is set 8, 14, or 20 dB higher than the value used to obtain the DAC curve: attenuation corrections may be needed. In Europe, the usual procedure is to increase the sensitivity setting until grass is just visible on the display. In the USA an amplitude level is usually specified and only defect indications which exceed that level (e.g. 20% DAC or 50% DAC) are recorded. The amplitude level to be used is specified in terms of a calibration block. The latter method produces a quicker, more reproducible test, but also a danger of missing indications from large but badly oriented flaws.

DGS (distance–gain–size) curves, already mentioned, and shown in Figs 4.44 and 4.45 are also used for sensitivity setting and for specifying the minimum size of a flaw which has been located. The DGS curve effectively relates the size of a circular, flat (ideal) reflector to the echo height obtained at different distances from the probe, and the gain setting. As no natural flaw is a perfect reflector, this gives the *minimum* size of the flaw. A DGS diagram cannot therefore be used to size defects unless these are considered to be planar reflectors, ideally oriented to the ultrasonic beam. The DGS diagram is theoretically determined and does not normally allow for attenuation effects. It is possible, on ultrasonic equipment, to introduce a variable gain control which corrects for the shape of a DGS curve and effectively produces the same echo height from the same flaw at different distances from the probe.

The DGS system does not therefore, strictly, evaluate the size of a flaw, but offers a standardised procedure for specifying the size of an echo obtained from any flaw which is located. The American Welding Society D1.1 Code follows a similar procedure of comparison with a known artificial reflector.

There are three techniques which attempt to assign a real size to a weld defect, and the validity of these methods and the attainable accuracy has been the subject of many research papers.

As part of the probe characteristics, the ultrasonic beam spread to some limiting value, usually -20 dB, should have been determined. If, therefore, the echo height is adjusted to a convenient value on the screen and the probe is traversed towards the weld until the signal height has fallen by 20 dB (i.e. to $\frac{1}{10}$ of its original value), the edge of the flaw will be intercepting the lower boundary of the ultrasonic beam (position A in Fig. 4.62): the probe-to-weld distance and the screen range are carefully measured. The probe is now traversed away from the weld, again until a 20 dB drop is obtained and similar measurements made (position B). If this data is now transferred to a flaw location slide-rule, or a large-scale diagram, the through-thickness dimension of the flaw can be obtained. This technique is widely known as the '20 dB drop' method.

The assumptions behind this procedure seem somewhat questionable. Firstly, the 20 dB isobar on the beam is not always easily definable because of secondary peaks. Secondly, if a defect is just touched by the 20 dB isobar, virtually the whole of the ultrasonic beam is unimpeded, and to obtain a drop to $\frac{1}{10}$-intensity from the flaw completely within the beam, some of the flaw must be inside the 20 dB isobar rather than just touching it. Fortunately, the change in intensity at the edge of the ultrasonic beam is relatively rapid, unless affected by side-lobes, and this will minimise the errors.

A very similar technique is known as the '6 dB drop'. The underlying assumption is that the echo signal displayed when the probe is aligned for maximum response will fall by half (-6 dB) when the axis of the beam is brought into line with the edge of the flaw. This method ought to be reasonably accurate if the size of the flaw is similar to, or larger than, the beam diameter, but is unlikely to be satisfactory for small flaws.

The third method is known as the 'maximum amplitude technique'. This takes into account that most flaws are facetted or rough, and do not present a single smooth surface to the ultrasonic beam: there is therefore a signal envelope rather than a single pulse. As the probe is scanned across the surface, there will be a series of maxima on individual pulses within the signal envelope. The surface distance of probe movement is measured, if necessary increasing the gain until the extreme facets of the flaw have been covered, and the range of probe movement on the surface, together with the timebase range, gives the size of the flaw.

All three methods require considerable skill to apply and the accuracy of the result will depend to a great extent on the nature of the flaw.

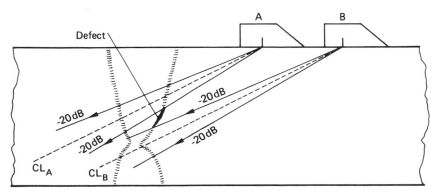

Fig. 4.62   Defect sizing by the intensity-drop technique (-20 dB)

In general these methods tend to underestimate flaw sizes and a series of recent research papers on flaw sizing accuracy have suggested that the likely error (standard deviation) on the through-thickness dimension is about 3 mm for the 20 dB drop method, 5 mm for the 6 dB drop, and 2.5 mm for the maximum amplitude method. Similar errors have been suggested for the flaw length measurement and there appears to be no correlation between absolute flaw size and accuracy of measurement. The implication of these errors is that the smallest flaw which can be measured with any accuracy is about 2.5 mm. With the TOFD method of flaw size measurement, which does not depend on the ultrasonic beam characteristics, greater accuracy in flaw sizing can be expected and a mean error (standard deviation) of 0.5–1 mm (2% of the specimen thickness) has been reported. The focussed probe technique (FET) has also been reported as measuring flaws with a standard deviation of 0.5–0.8 mm. As the TOFD technique is further developed to be a search technique as well as a sizing method, it seems likely that the other methods of weld flaw sizing may be less widely applied.

These errors in flaw sizing must not be confused with the defect detection probability studies, i.e. reliability of ultrasonic testing. In manual ultrasonic testing large variations in reliability have been reported due to variations in operator skill. Average probabilities of less than 50% have been reported compared with 80% on mechanised ultrasonic inspection.

The intensity of elastic wave scattering from a discontinuity is a function of direction, as well as other factors such as the size of the discontinuity, its surface state, and its wavelength, and in recent years formulae have been developed for the calculation of scattered intensity. If this intensity can be measured in terms of beam angle, the values might be used to determine the size and shape of internal defects.

A number of theories of elastic wave scattering have been proposed, and in practice some approximations are necessary. The Born approximation, originating in quantum mechanics theory, can be applied to scatterers of complex shape, and appears to be satisfactory when the wavelength is an order of magnitude larger than the scatterer, and also if the scatterer can be treated as elastic; however, the mathematics is complex, and present results appear to be a considerable way from dealing with the practical case of an incident pulse with a complex spectrum, and a real defect in metal.

The methods briefly described above for flaw location and flaw sizing in welds have been illustrated chiefly in terms of the inspection of a fusion butt-weld in a plate. On curved specimens and complex geometry welds, the procedures become progressively more difficult and require more care in determining the true direction and extent of the ultrasonic beam. A particularly complex case is that of butt- or fillet-welds in small-bore pipework, which are often counterbored. Contoured, miniature shear wave probes are usually used, both single and twin crystal, and a special calibration block in the form of a small tube with artificial holes is preferred to the standard calibration block.

On a curved surface, the gap between the metal and probe surface should not exceed 0.5 mm and the probe shoe should be adapted to enable the gap to be kept small. If the probe is so adapted, the probe index and probe angle must be redetermined. If shaped probe shoes are used, the beam spread is likely to be increased. A few examples of probe positioning on large fillet-welds are shown in Fig. 4.63.

Weld testing with shear wave probes is possible at temperatures up to at least 250 °C provided that the effect of temperature on sound velocity and attenuation is taken into consideration. The shear wave velocity, V_s, decreases with temperature, T, according to an empirical relationship:

$$V_s = -aT^2 - bT + c$$

where a, b and c are positive constants.

Attenuation changes in steel, with temperature up to 300 °C, have been found to be very small and can be neglected. In high-temperature weld testing, the specimen usually exhibits a temperature gradient and this causes changes in

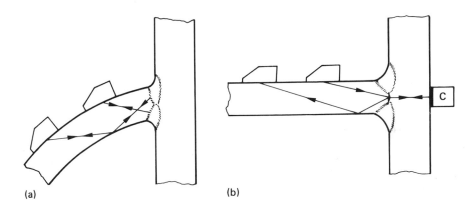

(a)   (b)

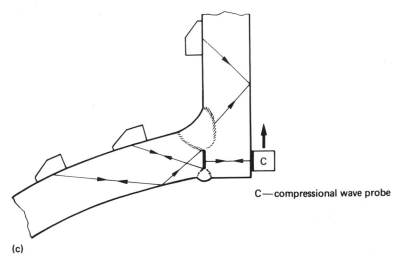

C—compressional wave probe

(c)

Fig. 4.63 Fillet-welds: probe positions for (a) full penetration weld, and (b) and (c) partial penetration welds

the angle of refraction of the beam. Metal probe wedges and glycerine couplant are usually used.

4.6.3   Welds in austenitic steel

Ultrasonic testing of welds in austenitic stainless steel is a much more difficult problem than weld testing in ferritic steels; with some alloys and thicknesses, it is at present impossible. The austenitic weld metal consists, in most cases, of large elongated anisotropic grains forming a fibrous structure with symmetry around the fibre axis. There are therefore a number of basic problems.

(1)   There is severe attenuation, particularly at high frequencies.
(2)   The anisotropic grain structure (dendrites) causes the ultrasonic beam to bend, particularly with shear waves: this means that ultrasonic propagation is a function of the grain orientation.
(3)   The material is noisy—there is a large amount of grass on the display, which can completely mask the pulses from flaws.

At the present time the problems are not fully understood: much depends on the welding procedures employed. Thus Juva and Lieto (1980) found few problems in testing 2–10 mm stainless steel welds with conventional techniques using both shear and compressional waves up to 6 MHz; 70° probes gave the best results. The material tested was 18% Cr, 11% Ni, 2.7% Mo steel. Likewise, Skorupa (1983) tested 19% Cr, 9% Ni austenitic steel welds in the thickness range 10–24 mm with both shear and compressional waves, the welds having been fabricated by a range of welding processes. Some investigators have found traces to be so noisy that all flaw echoes were lost; in other cases, large, spurious echoes were obtained, apparently from regions of the weld where no flaws existed.

On thicker welds, the main problem was thought at one time to be due to ultrasonic scatter, but more recent work has pin-pointed the problem as one of elastic anisotropy of the different grains. In welded austenitic steel, strong ⟨100⟩ fibre textures are produced with the ⟨100⟩ fibre axis parallel to columnar grain boundaries. The orientation of the fibre axis can vary over a large angle with small changes in the welding procedure. Ultrasonic propagation depends strongly on the orientation of the ultrasonic beam to the fibre axis. Also, the ultrasonic beam width varies with orientation, being smallest when the ultrasound is propagated in the direction of maximum velocity, when the signal-to-grass ratio is also a maximum. This optimum direction has been found to be at approximately 45° to the fibre axis. This beam skewing effect is a very serious problem (see Fig. 4.64); it leads to misjudgements of the position of a defect, and to apparently spurious signals. The ultrasonic inspection technique must therefore be planned in relation to the metallurgical structure of the weld, as well as to the weld geometry; further, the weld procedure and weld preparation need to be planned in terms of future ultrasonic inspection. Besides beam skewing and beam width variations, there can be ultrasound refraction at the weld fusion face, leading to additional beam deviation. In extreme cases there will be mode conversion as well as refraction.

Attenuation effects in austenitic steels are also not fully understood. The attenuation factor increases with average grain size, but is also a function of the (grain size)/wavelength ratio. However, steel samples with very different grain sizes can apparently have similar ultrasonic attenuation. There is still controversy over the effect on attenuation of the ferrite content, and some of the effects noted may be due to the presence of microcracks.

From the ultrasonic testing point of view, beam direction in relation to the grain fibre axis is all-important. Compressional waves are to be preferred to shear waves, as they give a higher signal-to-grass ratio as well as less beam skewing. Short-pulse probes should be used, as reducing either the beam spread or the pulse length reduces the volume from which scatter originates at any given moment. Similarly, focussed probes, or focussed probe pairs, will give an improvement in signal-to-grass ratio, by reducing beam spread. Defect sizing by probe movement is of course impossible if there are large fluctuations in attenuation as the probe is scanned.

In a typical butt-weld, there may not be uniform grain orientation, due to cooling effects at the fusion faces: also, in this region, rapid cooling can result in smaller grains. There are therefore, again, serious beam skewing effects which prevent a flaw being accurately located, even though it is apparently detected. The transmitted beam can be bifurcated, giving two maxima on each received signal.

A whole range of new ultrasonic techniques has been investigated in order to improve the ultrasonic inspection of austenitic steel welds. Some of these, such as SAFT (synthetic aperture focussing technique), will be discussed in Section 4.7, but of others, some can be briefly described here.

To improve the signal-to-noise ratio, amplitude modulation of the radiofrequency pulse exciting a wideband transducer has been proposed: this also requires special coupling to a tuned receiver circuit. Effectively, the transmitting probe is excited within a narrow bandwidth, using a pulse of a definite shape such as a sine or triangular wave. Noise improvements of up to 15 dB have been obtained.

Similarly, pulse compression techniques have been investigated. The received pulses are connected to a dispersive filter, so that the different frequencies are compressed in time at the filter output.

Signal averaging techniques have been used to enhance the defect signal and

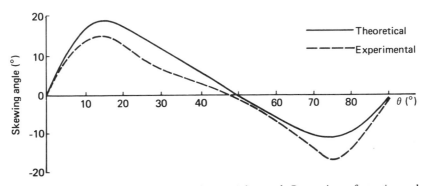

Fig. 4.64 Beam skewing in the inspection of 316 stainless steel. Comparison of experimental and theoretical results for compressional waves, where θ is the angle between normal to wavefront and columnar grain axis. Reproduced by permission of Nuclear Electric plc, from Coffey J M and Whittle M J, *Industrial Applications of Ultrasonic Testing*

suppress the noise. This can easily be done if the signals are digitised. Alternatively, if the probe is moved very slightly, the flaw echo is virtually unaffected, but the noise changes, thus producing spatial averaging (autocorrelation). Angular averaging by moving the probe over a small angle (±3°) has also been successful.

Multifrequency analysis of the received signal has been proposed, and as this technique has been suggested for a wide range of applications, it will be described in Section 4.7.

It has been suggested that for coarse-grained materials, a considerable improvement in signal-to-noise ratio can be obtained by further consideration of the signal output employed—the electrical excitation of the probe—rather than subsequent signal processing or the use of special probes. In the CS (controlled signals) technique a conventional probe is used, but the spectral distribution and centre frequency of the ultrasonic beam are adjusted to match the metallurgical structure of the particular material being tested. By choosing a frequency which is in the transition range between reflection and scattering from the flaws to be detected, and a pulse of 4–5 times the wavelength, it is claimed that good results can be obtained at a much lower frequency than is usually used. For example, good results in sizing intergranular stress corrosion cracking in stainless steel have been claimed for ultrasonic frequencies as low as 0.7 MHz. It is claimed that the frequency, f, at which there is a transition from reflection to scatter is given by

$$f = 0.5C/h$$

where C is the velocity and h is the defect height.

Adaptive learning networks (ALNs), using a computer 'trained' to search for special attributes of the flaw signal, have also been investigated. TOFD techniques have been found to be possible on austenitic welds, with, for deep cracks, an acceptable accuracy.

Ultrasonic inspection of austenitic steel welds is a difficult application at present. There appears to be no general rule and a procedure must be developed by practical trials on representative test-pieces containing artificial and natural flaws. Flaw sensitivity is likely to be inferior to that obtained on ferrite specimens of similar thickness. For thin welds, less than 15 mm, miniature shear-wave probes should be used; for thicker welds compression waves are necessary, and much attention must be paid to the welding procedure, the weld geometry and the ultrasonic beam direction, if spurious echoes and false flaw locations are to be avoided. Generally, short pulse, low frequency probes should be used to maximise depth resolution and minimise grain scatter, although there are problems in damping the probe crystals efficiently at low frequencies. One solution is to drive a transducer well below its natural resonance frequency using a waveform designed to produce the desired output pulse. This waveform will vary according to the grain structure.

Spatial averaging (averaging over several ultrasonic paths) can also improve flaw detectability. Rotational averaging with a multi-axis scanner has also given improved results.

In materials showing a high degree of scatter, frequency averaging has been shown to improve inspectability. To be successful, the flaw must be detectable over a range of test frequencies (Silk, 1987). The ultrasonic beam should be at

45°–50° to the columnar grain axis, which of course may not always be possible, or known, in practical applications.

Many of the problems in testing austenitic steel welds, outlined above, apply also to some degree to the ultrasonic inspection of welds in materials such as Inconel, other nickel–chrome alloys, and copper alloys, which are also coarse grained. However, ultrasonic testing has been applied in these fields only to a very small extent and experience is limited.

Ultrasonic testing of welds in aluminium is generally analogous to testing in fine-grained ferritic steels; ultrasonic attenuation is small and there are no special problems.

4.6.4   Castings inspection

Castings are made in an enormous range of sizes and materials. There can be large differences in the metallurgical structure, according to the size of the casting, the heat-treatment, and whether the testing is to be done before or after heat-treatment. Likewise, the surface condition of a casting varies widely with the nature of the moulding process used. Large sand-moulded castings usually have poor surfaces which need to be prepared for ultrasonic testing; they also frequently have headers and feeders, which can complicate access to the casting surface, so that ultrasonic testing may have to be deferred until part-machining is complete. Lastly, castings vary enormously in shape, from large, simple billets to small, highly complex precision castings for various industries. For complex castings, ultrasonic inspection may be on local, critical areas, because 100% coverage of the casting volume is virtually impossible to achieve. Much of the art of ultrasonic inspection in this field is in planning the probe positions for the desired beam directions. On many large steel castings, the elimination of all casting defects is not always possible. Some defects such as centreline shrinkage and isolated gas cavities may not be detrimental to the serviceability of the casting, but will produce ultrasonic responses, and sometimes mask other, more serious defects.

The main problems in ultrasonic casting inspection therefore are as follows.

(1)   In some steel castings, there is heavy attenuation and the ultrasonic path lengths are long, so that a high-output, large-diameter ultrasonic probe is necessary. (Compressional waves are preferred, but shear waves are also used.)
(2)   In large castings, the casting defects are likely to be extensive in area and the emphasis is on defect location and extent rather than the problems of sizing small defects.
(3)   Adequate coverage of a casting is a very important problem: flaws commonly occur at section changes, when the ultrasonic beam direction is difficult to control. Some defects, such as centreline shrinkage, can completely mask volumes of metal from ultrasonic inspection.

There are several standards on the ultrasonic inspection of casting, some of the best known being ASME/Section V/T-534, ASTM-SA-609, and BS:4080.

Sensitivity calibration, particularly on aluminium castings, quite often involves using a series of test blocks containing flat-bottomed holes of different depths. The ASTM has detailed sets of reference blocks for both steel and aluminium castings. Using a set of blocks with $\frac{1}{4}''$ diameter holes, at different distances from $1''$ to $10''$, the response curve against distance is drawn.

The attenuator reading to bring each of the readings up to 75% full–screen height is then called the amplitude reference line (ARL). Seven quality levels of castings are proposed, based on the total area of defect giving a signal above the ARL. This is a strictly empirical standard, as with long testing distances the area on the casting surface over which the probe is moved will not necessarily be the actual area of the discontinuity in the casting producing the signal. For use with angle probes, the ASME has proposed calibration on $\frac{1}{4}''$ and $\frac{3}{8}''$ side–drilled holes. Some operators use the loss of backwall echo as a measure of defect size.

With smaller castings, standard immersion techniques can be used, and it is possible to use a universal–movement probe head or robot under computer control, to move the probe over a pre–programmed path, determined for individual casting shapes.

In an intermediate, semi–automatic method which has been used, the position of a hand–held probe is plotted from the readings of two position-sensitive potentiometers attached to the probe. By this means, a C-scan presentation can be produced, which gives a 'picture' of the defects similar to a radiograph of the specimen. The results can be displayed on an $X–Y$ recorder or on a storage oscilloscope.

Austenitic steel castings produce the same difficulties as austenitic steel welds, in that coarse columnar grains can cause severe beam skewing and grain boundary scattering. The macrostructure of the steel depends critically on the chemical composition and may be isotropic or highly anisotropic (Gilroy, 1987).

4.6.5   Methods using Lamb waves

For high-speed testing of a plate, strip, or wire, where the thickness is of the order of a few wavelengths, there is considerable attraction in using Lamb waves. These are generated by having compressional waves incident on the specimen at appropriate angles, α, derived from

$$\sin \alpha = V_L / V_P$$

where V_L is the velocity of propagation of the incident wave and V_P is the phase velocity of the desired Lamb wave. (See Fig. 4.65). The plate specimen can be considered as a wave-guide.

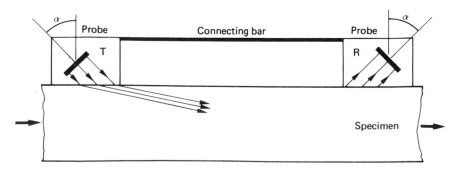

Fig. 4.65   Probe configuration to generate Lamb waves

The Lamb waves can be symmetrical or asymmetrical, and can be of different modes or mixed modes, and their velocity depends on the mode (see Section 4.2.3). In a symmetrical Lamb wave, the particles on the neutral axis have only longitudinal oscillations, whereas in the asymmetrical wave, the particles on the neutral axis vibrate transversely.

Lamb waves can also be generated and detected with an EMA (electromagnetic-acoustic) probe with the appropriate coil configuration.

The most widely applied technique is to monitor the transmission along a plate between two probes with a fixed spacing. For example, the quality of a spot or seam weld between two sheets can be monitored by the transmission along one sheet and through the spot weld (Fig. 4.66), and this can be done as on-line monitoring of the welding operation. Temperature effects at the weld

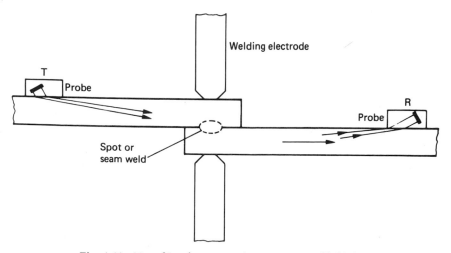

Fig. 4.66   Use of Lamb waves to inspect a spot-welded joint

nugget and mode conversions at the liquid weld/metal interface complicate the interpretation of the results. In wire specimens, the waves are usually known as rod waves and there is an advantage in generating them by magnetostriction, with a coil over the end, so that no surface contact is needed. Spirally rotating surface waves can also be generated in a wire with a pair of probes angled to the wire axis. Leaky Lamb waves have been used for testing composites. If the composite is immersed in a fluid, Lamb waves leak energy into the fluid at an angle determined by Snell's Law, and the leaky Lamb wave interferes with the specular wave component, with phase cancellation. By evaluating changes over a swept frequency range, very small changes in elastic properties due to flaws can be detected.

4.6.6   Tube and bar testing

There have been a number of very successful applications of ultrasound to the testing of wire, bar and tube specimens, and a few of these can be briefly described to illustrate the principles.

Figure 4.67 illustrates suitable probe configurations for tube testing. The

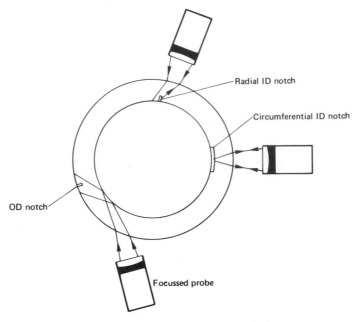

Fig. 4.67   Ultrasonic inspection of tubing

equipment will usually utilise several probes and operate in an immersion tank. The tubing can pass through the tank, or the probe assembly can be rotated around a fixed tube. The most suitable probe angle is usually chosen in terms of the diameter/thickness ratio of the tubing, as shown in Table 4.3.

Table 4.3   Shear wave angles for tube inspection

Angle (degrees)	Diameter/thickness
35	5
45	7
60	15
70	30
80	130

For the on-line testing of round bars or tubes at elevated temperatures, there is a considerable advantage in using EMA transducers (see Section 4.3). These enable a running clearance of up to 2 mm between the transducer and the metal surface. With a redesign of the EMA probe coil, using a zigzag flat grid, surface waves can be generated. The wires of the coil are spaced half a wavelength apart and surface waves are produced in two directions. The transducer is more dependent on lift-off distance than other forms of EMA transducer and it is preferable to use lower frequencies (0.3–0.5 MHz). EMA-generated surface waves have been used up to 1300 °C, but there is a strong dependence on temperature, the signal amplitude peaking at 700 °C, possibly due to a temperature gradient at right angles to the surface.

The use of Lamb waves (see Section 4.6.5) on thin-walled products, such as tubes, is an extension of surface wave testing. Lamb waves are vibrations on

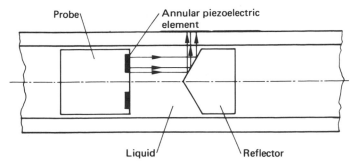

Fig. 4.68   Ultrasonic inspection of tubing using an internal probe and reflector

the whole thickness of a thin-wall tube or plate and can be generated and received by an EMA type of probe with a suitably-shaped coil. Lamb waves travel long distances with little attenuation; they can be sent radially, longitudinally or helically along a tube, according to the orientation of the generating grid.

Another interesting application of the ultrasonic testing of tubes is the measurement of wall thickness in tubes of 2.5–25.0 mm OD (outer diameter) using an internal probe and a metal mirror reflector (Fig. 4.68). The probe has an annular piezoelectric element, which produces a narrow ultrasonic beam which is reflected into the tube wall at right angles to the tube axis. Thickness variations of $\pm 10\,\mu m$ are claimed to be detectable. A focussed probe may be used to reduce the area covered at each position.

4.6.7   Bond testing

Dissimilar materials, such as rubber, composites, and metals, are frequently joined by a layer of adhesive, and there have been many attempts to measure the quality of the adhesive bond, by ultrasound.

The two materials concerned are likely to have different acoustic impedances, so that if a compressional wave is transmitted into one material perpendicular to the bonding face, some ultrasonic energy will be reflected back to the probe. If there is an air film indicating a local lack of adhesive on the interface, the amount of energy reflected will increase, so that it is relatively simple to sort out very bad bonds from good ones. The same argument applies to metal/metal interfaces, such as stainless steel cladding on a ferrous foundation. The usual technique is to use a combined transmitter and receiver probe, to eliminate the transmission surface pulse. ASME-SA-578 specifies calibration on a flat-bottomed $\frac{3}{8}''$ hole for this application.

Adhesive bond quality, however, is not so clear-cut, and it is doubtful how successful ultrasonic testing has been on detecting weak bonding due to poor cohesion, dirty surfaces, etc. To have any success, it is necessary to study each individual bonding procedure and the physical effects of variations in the bonding-procedure parameters. Ultrasonic testing must be carried out on test samples and a bond-strength calibration curve developed. This data is likely to apply to only one specific bonding procedure. There are alternative NDT techniques (e.g. thermography, holography, and acoustic emission) which appear to hold more promise than ultrasound for bond testing. It is possible to use Lamb waves for bond testing.

Theory suggests a relationship between the phase velocity of such waves and the quality of the interface, a higher phase velocity signifying a better adhesion, and there is some experimental evidence obtained from duralumin/epoxy interfaces to confirm this theory.

Another type of bond testing in which ultrasonics has had considerable success is on brazed joints in copper tubing. A through-transmission technique is used, with very narrow (1 mm diameter) water-jet coupling, which has a similar effect to a focussed probe. The joint is scanned on a helix pattern, mechanically, and displayed on a storage oscilloscope.

4.6.8   Composite material testing

Many complex material structures, such as graphite–epoxy and glass–epoxy laminates, honeycomb structure panels, and multilayer composites, are today used in the aerospace industry. Fibre-reinforced composites can contain a large range of potentially dangerous defects, and a wide range of NDT methods has been investigated (see also Chapter 8), as well as various ultrasonic techniques.

Ultrasound is used on many of these materials to detect lack of bonding, non–uniformity of structure, local defects, damaged honeycomb, etc. Most of the techniques are conventional and depend for their success on a thorough knowledge of the possible faults in the particular construction. The structures tend to be large and the inspection equipment is characterised by large scanning rigs under computer control, often in immersion tanks. The data is usually digitised, which enables a variety of methods of presentation and image processing to be used. Apart from pulse echo testing to reveal discrete defects, a favourite technique is to present the data in C-scan in pseudo-colour.

Ultrasonic velocity measurements have been used to monitor the polymerisation of the resin-curing process; ultrasonic attenuation measurements have been proposed to detect variations in composition, porosity content, and other anomalies. The spatial distribution of ultrasonic back scattering has been used to determine fibre orientation, by the application of a polar C-scan. Ultrasonic visualisation techniques have been used to display delaminations.

Spectrum analysis has been used to study defects, but is complicated by factors such as coupling and surface roughness, although deconvolution methods can be used to extract the relevant signal from the characteristic signal of the transducer, so reducing the number of factors involved.

One technique, called 'feature mapping', is closely analogous to pixel data display in optical imaging, in that the ultrasonic image is built up point-by-point over a chosen area, using a 15 MHz focussed probe. The 'feature' used need not be received pulse amplitude; it can be pulse duration, or pulse rise-time, or the ratio of the first positive and negative pulses on an unrectified display. Comparison of these various features is claimed to distinguish between porosity, delamination, and planar inclusions.

A special method which has been used for through-transmission on composite assemblies is the 'reflector plate' method (Fig. 4.69), with an immersion technique. As only one combined transmitter and receiver probe is necessary, mechanical coupling problems are eliminated.

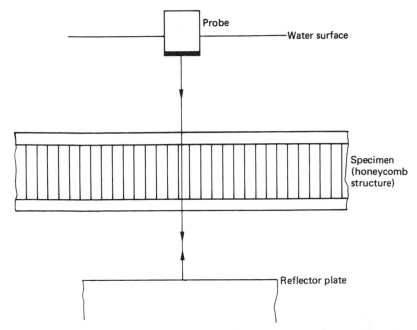

Fig. 4.69 Ultrasonic inspection of honeycomb structure using a plane reflector plate

4.6.9 Concrete testing

Concrete is not a precise material, being a complex of aggregate particles and voids in a matrix, but the desirability of a non-destructive test on both mature concrete and on newly-laid material is obvious. The detection and location of steel reinforcement bars is another application of NDT.

To obtain reasonable ultrasonic transmission in concrete it is necessary to use frequencies below 100 kHz, and this means long wavelengths. Taking a velocity of 4000 m/s, the wavelength at 50 kHz is 80 mm, and because the transducer is unlikely to be more than a few centimetres diameter, the ultrasonic beam has virtually no directional characteristics. At this wavelength path lengths up to several metres are possible. In addition, with such probes there is a large surface wave component and reflections of these from large aggregates in the concrete tends to obscure internal flaw echoes. For this reason a through-transmission technique is always preferable, whenever practicable, although for applications such as road surfaces a pulse echo must obviously be used.

In the past there have been many claims that ultrasonic velocity measurements, which depend on the overall elastic properties and the physical density of the concrete, can be related directly to concrete strength, particularly if concrete composition and moisture content can be controlled, but a number of surveys (Hillger, 1987, Teodoru, 1989 and Malhotra, 1984) have discredited this hypothesis. It is now suggested that a combination of pulse velocity measurement (V), plus ultrasonic attenuation (A), plus rebound index (R)

with a low velocity hammer is necessary to determine the compressive strength, F_c (N/mm^2), according to

$$F_c = a \cdot \exp(bV - cA) \cdot R^d$$

where a, b, c, d are constants. Teodoru (1989) has published nomograms for F_c for values less than 43 N/mm^2.

The practical problems are to know the ultrasonic path length and to measure the transit-time of a pulse. As the changes in pulse velocity are relatively small, the transit-time needs to be measured to within ±1% for pulse velocity measurements and this usually means using two probes on opposite sides, rather than a one-side-only technique. In making such measurements, it is desirable to avoid embedded steel reinforcement bars, although correction factors have been proposed for cases where these cannot be avoided (BS:4408:Part 5). A correction factor, K, relating concrete pulse velocity, V_c, to the measured value, V_m, when a bar is in line with the transducers, is given by

$$K = 1 - L_s[L(1-v)]^{-1}$$
$$V_c = KV_m$$

where L is the total path length, L_s is the path length in steel, and v is the ratio [velocity in concrete (V_c)]/[velocity in steel (V_s)].

Even when a bar is offset from the line between the transducers it can affect the ultrasonic velocity and a correction needs to be made (BS:1881:Part 203).

If the bar is transverse to the pulse path, a similar approach can be adopted, but the effective velocity of a pulse crossing the diameter of a bar has been found to be only about half the expected value, for so far unexplained reasons, so that bars of less than 20 mm diameter can usually be neglected.

4.7   SPECIAL TECHNIQUES

Ultrasonic non-destructive testing is a developing subject, and is characterised by there being a very large number of possible ultrasonic waveforms and also by the large amount of data available in any received pulses, which is not utilised.

4.7.1   Ultrasonic spectroscopy

It is relatively easy to analyse an ultrasonic pulse into a frequency spectrum—ultrasonic spectroscopy. Methods of using this data have, however, not yet had much practical application.

The frequency spectrum of an ultrasonic pulse can be easily obtained either by electronic methods or by digital methods of Fourier analysis and the spectrum should contain information on the nature of the reflecting target—the defect. The pulse can be represented by the sum of a series of waves of different amplitudes and different frequencies: a short pulse will have a wide frequency spectrum and vice versa.

This method of analysis can also be used to study the output pulse from various types of transducer, but as the spectrum will be affected by the coupling to the specimen, as well as by the constructional details of individual probes, it is difficult to see much practical use for such information. It has been claimed that defect geometry can be studied from the spectrum of the received pulse, and it has been shown that the spectra from a spherical cavity, an in-line crack, and an inclined crack are quite different. However, the spectral signatures of different defects must depend on the spectrum of the input pulse and at present it is not possible to standardise this; also, the defect spectrum will be affected by any attenuation of the signal in the material.

The technique of spectrum analysis has produced considerable information on transducers. It has shown that the nominal resonance peak frequency of a transducer can be far from the actual measured response and that the method of driving the transducers is also a critical factor.

A short pulse of ultrasound is a form of coherent radiation, so that in a thin-wall specimen which produces front and backwall echoes, the two reflected pulses show phase differences and they can interfere coherently. If the pulse contains a wide band of frequencies, interference maxima and minima can occur at particular frequencies and these can be related to the specimen thickness.

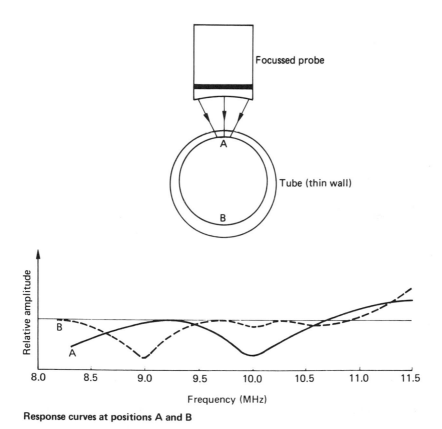

Fig. 4.70 Very-thin-wall tube inspection: frequency spectrum variations due to wall-thickness variations (10% change in a 0.3 mm wall thickness)

For example, if a 10 MHz focussed probe is used on a 0.3 mm metal tube, there is a strong minimum at 9 MHz, f_{min} (Fig. 4.70), given by

$$f_{min} = V/2T$$

where V is the ultrasonic velocity and T is the tube-wall thickness.

As the tube-wall thickness changes, a change in frequency, Δf, can be related to the thickness change of ΔT

$$\Delta T = T_A(\Delta f/f_{min})$$

where T_A is the nominal thickness.

It is not therefore necessary to know the value of V to determine ΔT, and probe alignment is not critical.

4.7.2   Acoustic holography

Optical holography is now well known as a method of storing three-dimensional data on a two-dimensional recording surface, and as an ultrasonic pulse is a form of coherent radiation, it is possible to apply similar techniques to ultrasound. It must, however, be remembered that ultrasound as used in NDT has a long wavelength, of the same magnitude as the size of the defects to be detected. The ultrasonic aperture size is also only a few wavelengths, so longitudinal image resolution is likely to be poor.

The principle of linear acoustic holography is that the ultrasonic transducer is scanned along a line and produces a wide sound field in the plane of the scan line. The received echoes are multiplied with two reference signals, one 0° and one 90° shifted against the transmitted pulse, so producing a real and imaginary part of the received echo at a number of sampling points along the aperture. The data is stored in a computer and by using quadratic phase functions, the computer defines a reconstruction plane and an observation angle, and calculates the ultrasound field intensity at the region of a defect. From the profile of this ultrasound intensity field, the defect size can be derived. The main disadvantage of acoustic holography is the poor resolution in the direction of the ultrasound field, but this can be overcome to some extent with a focussing probe, or by multifrequency holography, or by synthetic aperture focussing techniques (see Section 4.7.3).

Early work on acoustic holography used an immersion system with two ultrasonic transducers with overlapping ultrasound fields at the specimen, one transducer producing the reference beam. The interfering fields produced a static surface displacement on the liquid surface, which was optically illuminated with a laser beam to produce a hologram of the specimen. Later, an electronic reference beam was used, but both these systems appear to have been superseded by computer reconstruction techniques with synthetic aperture scanning.

4.7.3   Synthetic aperture focussing techniques (SAFT)

Various methods such as ALOK (Section 4.5.6) have been proposed for collecting and processing data from a scanned transducer, so as to simulate a much larger transducer having much better resolution and signal-to-noise ratio. SAFT, applied to an ultrasonic transducer, is derived in principle from synthetic aperture radar systems. In theory, it is possible to take into account irregular refractive interfaces in the specimen and complex angular scattering by defects.

SAFT uses pulsed ultrasound, records both wave amplitude and phase, and does not need a reference function. The data from a series of probe positions is stored, after digitisation (Fig. 4.71). The time of flight from the probe to the defect, for different probe positions, is corrected by a time-based algorithm; depth of field effects can be overcome by a focussing algorithm and the uneven nature of the material interfaces by a surface-phase-correction algorithm; noise is reduced by averaging a series of pulses from each probe position. The practical aperture size, which is the size of the theoretical large transducer,

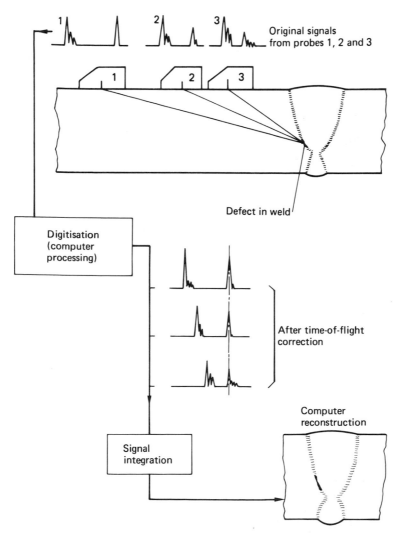

Fig. 4.71   Synthetic aperture focussing technique (SAFT): only three probe positions are shown for simplicity; after digitisation and time-of-flight correction, the signals are integrated

cannot of course be infinitely large, as the ultrasonic beam from the probe positioned at the extremities of the aperture must still reach the flaw.

A depth resolution of 0.4 mm with 1.4 mm wavelength has been claimed, but flaws with lateral dimensions less than λ cannot be sized. With parallel-processing, real-time imaging using SAFT techniques has become possible (Ozaki, 1987).

4.7.4   Ultrasonic imaging systems

An ultrasonic 'image' of a specimen can be obtained by the C-scan display already described, by scanning a probe over the surface, usually in an immersion tank, on a regular raster. However, such mechanical scanning is slow, and the transverse resolution is severely limited by the beam diameter. If a focussed ultrasonic probe is used, the field width is smaller, and transverse resolution is improved, but mechanical scanning times will be still longer. Electronic scanning is needed, together with a multielement linear array of transducers, to speed up the process.

Some time-delays must be built into the transducer elements so that a signal from one point in the specimen reaches all the elements at the same moment. Such multi-array systems have found widespread use for ultrasonic imaging in the medical field, but not yet for industrial applications. The scanning can be mechanical, or electronic using a steered array (phased array).

In a phased array, the element spacing must be less than half the wavelength to avoid spurious diffraction lobes, so there are difficult fabrication problems; however, there is also a useful flexibility between frame rate, line density, and field of view. Systems working between 3 and 15 MHz are now in use.

A whole range of equipment has been proposed for direct ultrasonic imaging in metals, from television-type camera tubes with a faceplate sensitive to ultrasound, through the use of ultrasonic lenses, to computer-reconstructed images. Ultrasonic lenses can be made of solid materials, or with liquids inside a thin, shaped skin. Lens design is analogous to optical lens design, with appropriate values of refractive indices, based on the ultrasonic velocities in the different materials.

Several methods have been used to make a spatial ultrasonic image visible, as follows.

(1)   Liquid surface levitation (as mentioned briefly for acoustic holography systems).
(2)   Elasto-optical effects: in suitable transparent solids, polarised light can make ultrasonic waves visible, by becoming optically double-refracting under stress.
(3)   Schlieren methods: the change in optical refractive index, when a transparent material is subjected to a pressure change, can be used to make visible an ultrasonic field.
(4)   Nematic liquid crystals can be stimulated by an electric field to become turbid and scatter light, and ultrasonic waves can produce the same effect.
(5)   Diffraction effects from the interaction of a light beam and an ultrasonic beam in a liquid: the latter can be regarded as an optical grating with changes in

refractive index along the grating. If the Bragg diffraction condition is fulfilled,

$$2 \sin \alpha = \lambda/\Delta$$

where α is the beam angle, λ is the optical wavelength, and Δ is the acoustic wavelength, the diffracted light carries an optical representation of the acoustic field.

Direct image visualisation techniques are attractive in that they produce instantaneous image displays, and equipment such as the radial–sector scanner, with a scan along radial lines extending from the centre of the transducer array, has been very successful in medical ultrasonic applications. The delay–line system is complex and is only possible for a small number of elements on the transducer, so that as analogue/digital converters get faster and cheaper, digital computer techniques seem likely to be used for the next generation of imagers, as already described under synthetic aperture systems (see Section 4.7.3).

The need to examine rapidly moving objects, such as the human heart, is not present in industrial non-destructive testing applictions, hence longer image build-up times can be tolerated. A present, most electronically-scanned ultrasonic systems for medical applications use a much smaller number of lines to form the image than a television raster, so that even with interpolation routines the image appears crude. For imaging, some of the problems in constructing transducer arrays can be eliminated by using a plastic sheet piezoelectric material PVDF.

4.7.5   Ultrasonic attenuation measurements

Recent technical developments have simplified the problems of accurate measurement of ultrasonic attenuation in metals over a wide range of frequencies and it seems possible that attenuation data might be used to predict material properties which are relevant to strength calculations.

A broadband, high-frequency transducer, driven by a fast-rise-time pulser is coupled to a specimen by a quartz buffer rod. The specimen thickness is normally small (1–20 mm), parallel-sided, and considerably wider than the transducer. Several return echoes from the back of the specimen are captured and processed, to correct for diffraction losses and reflections at the interfaces. Curves relating frequency, f, mean grain size, $\bar{D}$, and attenuation, α, can then be plotted. Papadakis (1984) has proposed:

$$\alpha = B_1 \bar{D}^3 A^2 f^4 \qquad \lambda > 2\pi\bar{D}$$

$$\alpha = B_2 \bar{D} A^2 f^2 \qquad \lambda < 2\pi\bar{D}$$

$$\alpha = B_3/\bar{D} \qquad \lambda \ll 2\pi\bar{D}$$

where α—attenutation, A—elastic anistropy of a single grain, f—frequency, D—mean grain diameter, B_1, B_2, B_3 are constants.

With low frequencies ($\lambda \ll \bar{D}$), it appears that attenuation is mostly due to geometric scattering, in the sense that the whole grain acts as the scattering unit. In this region, α is proportional to f^4. At very high frequencies ($\lambda \ll \bar{D}$), scattering is a diffusion process and is independent of frequency, and in the intermediate range there is 'phase scattering' due to the phase differences on the initially plane wavefront caused by the large number of paths associated with each elementary portion of the wavefront. In this region, α is proportional to f^2.

This data has been obtained largely from work in pure materials and while it is also valid for polycrystalline-alloyed materials, there is an additional attenuation effect due to dislocations, which may eventually make it possible to relate ultrasonic attenuation measurements to fatigue strength properties. At the least, ultrasonic attenuation can be used together with ultrasonic velocity measurements to monitor material properties, but at present there seems to be insufficient theory on the interaction of ultrasonic stress waves, material microstructure, and mechanical failure in metals.

There have been claims for empirical correlations between ultrasonic attenuation and the impact strength of certain steels, between attenuation and fracture toughness, and between compressional wave velocity and the tensile strength of cast iron. Using a single sample of stainless steel in which the fracture toughness could be varied with temperature, it was found that the correlation between ultrasonic attenuation and fracture toughness, K_{1c}, was not good, the variation in K_{1c} being much larger than the variation in attenuation. The reason was thought to be variation in other material properties, such as yield stress, also affecting K_{1c}, whereas ultrasonic attenuation is mostly dependent on grain size.

4.7.6   Stress measurement

There is a need for a portable NDT method capable of measuring stress within specimens, in situ, and since ultrasonic velocity is dependent on stress, the use of most types of ultrasonic wave has been investigated. The change in velocity is likely to be of the order of only 0.3%, so the method is unlikely to be easy experimentally. In addition, change in velocity varies with different materials, but the increased ability to time the arrival of ultrasonic pulses accurately (± 1 ns) has made this a feasible technique with some practical applications, such as the measurement of axial loads in steel bolts and residual stress measurement. The velocity of ultrasound also varies with temperature, so that specimen temperatures must be accurately controlled. The real limitation to the technique is that in many materials the ultrasonic pulse becomes distorted and so accuracy of measurement is reduced.

The acousto–elastic effect with Rayleigh surface waves has been used to measure residual stress in stainless steel pipes. EMA transducers are used to eliminate coupling problems and held on a rigid framework to improve repeatability. 1.5 MHz frequency is used and a calibration curve is developed on a test sample of the pipe using strain gauges to measure the stress. The Rayleigh wave velocity at zero stress is unknown and is affected by microstructural variations, so the main use of the technique is for relative rather than absolute measurements.

Improved results have been claimed from the use of shear waves polarised parallel to the surface, and it has been suggested that these waves are insensitive to acoustic anisotropy produced by texture effects in steel, which normally tends to mask the effects of stress change. Other workers have found advantage in using sub-surface longitudinal waves produced by critical refraction.

Another technique which avoids this problem is to measure the phase difference between two-tone bursts, by changing the frequency to keep the phase difference constant. Small specimens are used, in a water bath, and the pulses received from the front and back surfaces overlap.

The presence of stress rotates the plane of polarisation of polarised shear waves and there is some correlation between the angle of rotation and the magnitude of the stress. Measurement of this rotation can therefore be used to measure internal stress in a solid averaged over the volume of material traversed by the ultrasonic beam.

A more advanced technique for measuring residual stress is to use shear waves polarised in two mutually perpendicular directions. These waves have slightly different velocities and so interfere, so that as the transducer is rotated, the interference vanishes when the polarising planes are parallel and perpendicular to the stress axis. Once the axis is known, the actual stress can be computed from the velocities.

4.7.7   Critical angle reflectivity

If the reflectivity from a liquid/solid interface is measured, there is a very sharp minimum reflectivity at the Rayleigh critical angle (Fig. 4.72). From knowledge of this angle, the Rayleigh and Lamb wave velocities can be determined, and the shape of the minimum is affected by the attenuating properties of the surface layer of the solid, or of the interface properties between two different materials. The measurement is made with an ultrasonic goniometer and for most metals the value of critical angle can be determined to ± 1 minute of arc, so that the surface wave velocity can be measured to ± 1 m s^{-1}. The technique can therefore be used to measure the tensile strength of grey cast iron, to measure cold-work conditions on austenitic steels, to study heat-treatment or texture, and to measure the thickness of thin surface layers, such as the cementation thickness.

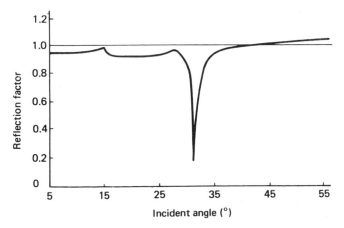

Fig. 4.72   Critical angle reflectivity at a water/stainless steel interface showing a sharp minimum reflectivity at 31°

The ultrasonic frequency at which minimum reflection occurs can be related to ultrasonic attenuation and grain size, but the relationships are exceedingly complex and few practical applications have so far been found for this measurement.

4.7.8   Ultrasonic microscope

A device using very-high-frequency ultrasonic waves (400 MHz to 2 GHz) has been proposed by a number of workers for microscopic examinations of very thin specimens. At these frequencies, the ultrasonic wavelength is a few micrometres in water, and at 400 MHz, a resolution of 3 μm is claimed.

One design of ultrasonic microscope is shown in Fig. 4.73. This uses a single surface refracting lens, usually made of sapphire, which, with water-coupling, gives a speed ratio in the two materials of 7.5:1, so that large-angle apertures with virtually no spherical aberration are possible. However, there are problems in separating the acoustic echo from the object from reverberations within the lens, and it is usual to work in the pulse mode. There is an upper limit on the focal length of the lens, due to the attenuation of the coupling fluid, and at 1 GHz ($\lambda = 1.5 \mu$m in water), the water path must be less than 100 μm. Even when imaging beneath the surface of a specimen, with the lens almost touching, the concave nature of the lens produces a residual water path and hence attenuation losses.

An alternative design of acoustic lens is shown in Fig. 4.74. Rays from the transducer are reflected from the curved surface and refracted out of the front surface to a focus: the front surface is flat, so that the fluid path can be negligibly small. The lens material and the lens profile must obviously be such that the times of flight for all the rays are equal, and if the object material has the same sound velocity as the lens material, the shape will be parabolic. The

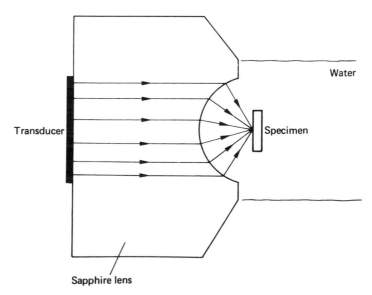

Fig. 4.73   Design of ultrasonic microscope with a sapphire focussing lens

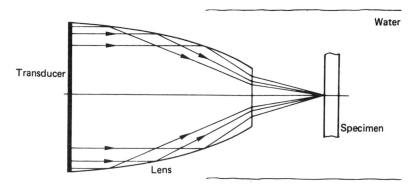

Fig. 4.74   Proposed design of an ultrasonic microscope using internal reflection in a parabolic lens

central portion of the lens is not irradiated, so there can be large side-lobes. The main applications are the imaging of detail below the surface of small opaque specimens to detect internal cracks, voids, and laminations, such as, for example, delaminations at different depths through the thickness of a 'chip' on a wafer. Penetration can be up to 10 μm.

In this type of ultrasonic microscope, the specimen is scanned mechanically and different layers of the specimen are imaged by axial movement. Liquid metal couplants can reduce interface losses and allow higher frequencies to be used with corresponding higher resolution.

A scanning acoustic microscope (SAM), operating at 50 MHz and using water as a couplant has been used to examine plasma-sprayed ceramic specimens, bulk ceramics, brazed and diffusion-bonded joints.

A scanning photoacoustic microscope (SPAM) is also possible, using a laser spot scanning over the surface to produce an acoustic signal, whose magnitude depends on the optical and thermal properties of the specimen. The advantage of laser generation is that it is non-contact.

Further reading

Höller P (ed), *New Procedures in NDT*, Springer-Verlag, Berlin, 1983
Handbook of the Ultrasonic Examination of Welds, The International Institute of Welding, 1977
Krautkrämer J and Krautkrämer H, *Ultrasonic Testing of Materials* (3E), Spring-Verlag, Berlin and New York, 1983. ISBN038711733
Procedures and Recommendations for the Ultrasonic Testing of Butt-Welds (2E), The Welding Institute, Cambridge, 1971
Silk M G, *Ultrasonic Transducers for NDT*, Adam Hilger, Bristol, 1984
Silvius H S (Jr), *Advanced Ultrasonic Testing Systems*, NT1AC-77-1, San Antonio, USA, 1976
Thompson D O and Chimenti D A (eds), *Reviews of Progress in Quantitative NDE*, Volumes 1, 2, 3 and 3A, Plenum Press, New York, 1982–84

References

BS:3923:1986, *Ultrasonic testing of welds*, British Standards Institution, London
BS:1881: *Testing Concrete*, BSI, London
Part 201:1986, *Guide to use of ND methods of test for hardened concrete*

Part 202:1986, *Recommendations for surface hardness testing by rebound hammer*
Part 203:1986, *Recommendations for measurement of velocity of ultrasonic pulses in concrete*
Bonami D and De Vadder D, *Eng Fracture Mech*, **23(5)**, 913, 1986
DD:174:1988, *Guide to the calibration and setting up of the ultrasonic TOFD technique for the location and sizing of flaws*, BSI, London
De Vadder D and Lhemery A, *Proc 12th World Conf NDT (Amsterdam)*, **2**, 1237, Elsevier Press, 1989
Dieulesaint E and Roger D, *Elastic waves in Solids*, John Wiley, 1982
Gilroy K S, *Proc 4th Europ Conf NDT (London)*, **3**, 1634, Pergamon Press, Oxford, 1988
Harker A H, *Elastic Waves in Solids, with Applications to the NDT of Pipelines*, Adam Hilger, 1988
Hillger W, *Proc 4th Europ Conf NDT (London)*, **2**, 1003, Pergamon Press, 1988
Johnson A, DRUID—a colour graphics ultrasonic imaging device, *Brit J NDT*, **26(4)**, 1984
Keller J B, Diffraction by an aperture, *J Appl Phys*, **28**, 426, 1957
Le Brun A and Pons F, Non-contact ultrasonic testing, *Proc 4th Europ Conf NDT (London)*, **3**, 1593, Pergamon Press, 1988
Malhotra V M, *In-situ NDT of Concrete*, Amer Concrete Inst Report, SP-82-831, 1984
Nussmuller E, *Proc 4th Europ Conf NDT (London)*, **2**, 1174, Pergamon Press, 1988
Ozaki, Y, *Proc 4th Europ Conf NDT (London)*, **4**, 2489, Pergamon Press, 1988
Papadakis E P, Physical acoustics and microstructure of iron alloys, *Int Metal Rev*, **29(1)**, 1984
Silk M G, Chapter 2 of *Research Techniques in NDT*, Volume 3, Sharpe R S (ed), Academic Press, 1977
Silk M G, *Proc 4th Europ Conf NDT (London)*, **3**, 1647, Pergamon Press, 1988
Temple J A G, Time of flight inspection: theory, *Nuclear Energy*, **22(5)**, 335, 1983
Temple J A G, Predicted ultrasonic responses for pulse-echo indications, *Brit J NDT*, **28(3)**, 145, 1986
Teodoru, G, The NDT of concrete, *Proc 12th World Conf NDT (Amsterdam)*, **2**, 1237, Elsevier Press, 1989

5

Magnetic methods

5.1 BASIC PRINCIPLES

Magnetic effects are explained by the concept of electromagnetic fields, which can be imagined as lines of magnetic force extending through space. The laws of electromagnetism were formulated by Maxwell, about 1873.

When a specimen is magnetised, the magnetic lines of force (the magnetic flux) are predominantly inside the ferromagnetic material. If, however, there is a surface-breaking flaw, or a subsurface flaw, the field is distorted, causing local magnetic flux leakage fields. Effectively, the flaw causes a sudden local change in permeability, μ.

Magnetic methods of NDT depend on detecting this local magnetic flux leakage field. The production of mathematical models of the flux leakage field due to various defects is possible, but some early attempts have used limiting conditions which are not realistic. More recently, using large computers, better models have resulted and rapid progress in this modelling field can be expected. The starting point is Maxwell's equations and finite element analysis. Large matrix equations are produced and these can be solved by iterative methods, or by conjugate gradient methods. The results show that the size of the leakage field is dependent on a large number of factors, most of which are irrelevant so far as present-day magnetic methods of NDT are concerned.

The published literature on magnetic methods is based largely on experimental empirical data, and by far the most widely used magnetic method of inspection is the 'magnetic particle flaw detection' technique.

The specimen, which is ferromagnetic, is magnetised by suitable methods, and flaws which break the surface, or are just subsurface, distort the magnetic field, causing local flux leakage fields. Finely-divided ferromagnetic particles are applied to the surface of the magnetised specimen and are attracted by the flux leakage field; they accumulate at the site of the flaw, producing a build-up of particles which can be seen by the eye, thus enabling the flaw to be detected.

The method is only applicable to materials which can be strongly magnetised—ferritic steels and irons (but not all steels). The magnetic particles can be applied as a powder, or more commonly as a liquid suspension, usually known as 'magnetic ink'. The accumulation of particles can make a very fine flaw such as a very narrow crack easily visible, even though it cannot be seen by the unaided eye. To be detected, linear flaws, such as cracks, must be favourably oriented in relation to the direction of the magnetic field, and also, the colour of the magnetic particles should be in good contrast to the colour of the surface of the specimen.

Two quantities are necessary to describe magnetic fields—the magnetic field intensity, H, measured in ampères per metre (A m^{-1}), and the magnetic induction, B (also known as the magnetic flux density), measured in teslas (T). Both are vector quantities, i.e. they have magnitude and direction. To a first approximation, H and B are related by

$$B = \mu_0(H + M) \quad \text{or} \quad B = \mu_r\mu_0H$$

where M is the intensity of magnetisation in the material and μ_0 is the permeability of free space ($4\pi \times 10^{-7}$ henries m^{-1}). μ_r is the 'relative permeability', sometimes simply called the permeability.

In air and non-magnetic materials, the magnitude of the intensity of magnetisation, $M \to 0$, so $\mu \to 1$; in magnetic materials, M is very large. In some standards, the required degree of magnetisation for magnetic particle flaw detection is specified in terms of the magnitude of the magnetic field intensity, H (e.g. $H > 2400$ A m^{-1}), but this may be misleading and it is better practice to specify a value of B (e.g. 0.72 T).

For example, BS:6072:1981 specifies a value of 0.72 tesla and assumes that the permeability will be greater than 240. Many engineering steels have a much higher permeability and a magnetising field of 2400 A m^{-1} will therefore give values greater than 0.72 T.

The tesla (T) is the SI unit of magnetic flux density (1 Weber per square metre). For conversion from older units $1\,\text{T} = 10^4$ gauss.

In a closed specimen, such as a ring, if there are no surface flaws, the flux density is confined within the specimen and a very low magnetic field exists in the air. If, however, there is a surface flaw in the ring, the flux lines have to cross the gap, and in the air gap they spread out, producing a leakage flux, which is the basis of magnetic flaw detection. If the break in the ring (the flaw) was parallel to the lines of magnetic force, there would be little or no flux leakage. The leakage flux is therefore largest for flaws which lie at right angles to the lines of magnetic force: it is generally possible to detect flaws which lie up to about $\pm 45°$ to this direction. If the flaws do not reach the specimen surface (i.e. in the case of subsurface flaws), the leakage flux due to the flaw will be much reduced and will be spread over a larger area. Internal flaws are therefore not likely to be detected unless a very high value of magnetising field is used and other conditions (surface state, etc.) are ideal.

5.1.1   Eddy currents

Magnetic fields which vary with time induce electric currents in conducting materials. In solid materials the induced currents are called eddy currents or Foucault currents. These eddy currents set up secondary magnetic fields which oppose the original magnetic field in direction, so that the magnetic field inside the specimen decreases in intensity with distance below the surface. The resulting effect is as if the electrical and magnetic fields are confined to a thin layer close to the surface—the *skin effect*.

The standard depth of penetration, δ, is defined as the depth at which the intensity of the induced eddy currents is reduced to $(1/e)\%$ of its surface value, where $e = 2.718$ (i.e. to 36.8%),

$$\delta = (\pi f \mu_r \mu_0 \sigma)^{-1/2}$$

where f is the frequency of the a.c. field and σ is the electrical conductivity.

If an a.c. current is used to magnetise a specimen, the skin effect concentrates the magnetic flux close to the surface, and prevents the detection of sub-surface or deeper cracks. In a medium–carbon ferritic steel, for $f = 50\,\text{Hz}$, $\delta = 1\,\text{mm}$. Thus the inner surfaces of tubes, etc. cannot be magnetised by external a.c. fields, and alternative magnetising methods, such as the threaded bar (see Section 5.1.3), have to be used. It is sometimes suggested that the use of a.c. concentrates the induced field near the surface, but there is no theory to justify this. The increased sensitivity with MPI using a.c. magnetisation is due to an enhanced interaction between the magnetic particles and the a.c. leakage field.

5.1.2 Magnetisation: hysteresis

The properties of magnetic materials can be described by a graph of B against H (Fig. 5.1). If M is very large, B represents the magnetisation to a close approximation. At point O, both B and H are zero—the demagnetised state. As a magnetising field is applied, the value of B increases along OAM to a saturation value; if then the magnetic field is reduced, the magnetisation falls, but along curve MBCN. On remagnetisation, the value of B follows a new curve NPQM. This closed loop effect is known as *hysteresis*. The value of

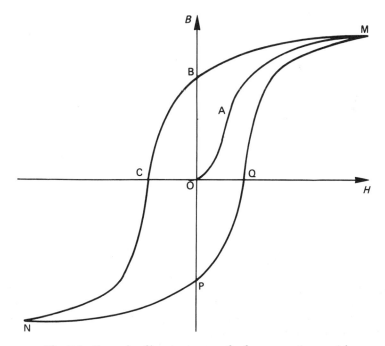

Fig. 5.1 Example of hysteresis curve for ferromagnetic material

magnetisation at B (zero magnetising field) is known as the 'remanent magnetisation' or 'remanence'. To remove the magnetisation, a negative magnetic field must be applied—the 'coercive force' or 'coercivity'.

The shape of the hysteresis loop is very different for different steels, although the value of magnetisation at saturation (point M) is approximately 2 T (1.6–2.1 T) for most steels and cast irons. In practice, saturation is rarely reached, due to the flattening of the B–H curve for high values of H. The field intensity to produce a value of $B = 0.72$ T can vary from 200–4000 A m^{-1} with different materials, so that in magnetic particle flaw detection it is necessary to choose suitable operating values to achieve optimum results. BS:6072 specifically chooses the value of $B = 0.72$ T as the value of magnetic induction required for flaw detection, based on an assessment of a wide range of magnetic materials, which showed that this value of magnetic induction was achieved at field intensity values less than 2400 A m^{-1} for nearly all materials. Some authorities, however, have recommended a higher value of B, for example $B = 1.1$ T, and where the recommendation is based on magnetic field strength, 3600–4800 A m^{-1} has also been proposed, but the practical value to be used must depend to some extent on the specimen surface roughness. Over-magnetisation on a rough surface can lead to 'furring' and an increase in background image noise. For specimens with a high surface finish, such as aerospace components, a value of H as high as 9000 A m^{-1} has been proposed. Flux density *in the specimen* is a difficult quantity to measure, and it is usual to depend on indirect measurements, or on the value of H. Hall-effect probes can measure the leakage magnetic flux close to the surface and values of 1–6 mT have been suggested as desirable for optimum detection of a small crack. Qualitative 'flux indicators' containing an artificial flaw have been devised which provide useful practical test indicators.

5.1.3   Magnetising techniques

Specimens to be tested can be examined overall or locally. It is not necessary for a specimen to be magnetised throughout its volume; local areas can be magnetised.

The methods of magnetisation can be broadly grouped as:

(1)   current flow techniques,
(2)   magnetic flow techniques,
(3)   induced current flow techniques,

and the choice between these will depend on the shape and size of the specimen, the expected direction of critical flaws, and whether an overall or a local examination is to be performed.

Residual magnetism

If a specimen is magnetised, and the magnetising field removed, it remains magnetised—point B in Fig. 5.1.

This residual magnetism can be used to carry out magnetic particle flaw detection. It is only in recent years that a magnetic flux value of at least 0.72 T has been specified and it is important to emphasise that there is no sudden cutoff value, below which magnetic particle flaw detection does not work.

There will be a gradual diminution in flaw sensitivity as lower magnetic fluxes are used, but even with low magnetic fluxes, flaws can be detected, and, in the past, much magnetic crack detection has been performed with residual magnetism. In fact, with many materials, the remanence value will be around 0.7 T. A more important point in using residual magnetism is that as the magnetising field is reduced to zero, the residual flux is not necessarily the value at point B, and its actual value depends on the shape of the specimen, as well as on the material characteristics. If the hysteresis curve is a wide loop, something approaching a square shape, a high value of residual magnetism is probable. This is usually the case with ring-shaped specimens, rather than with short, thick specimens.

If residual magnetism is to be used, the magnetising current should be direct current or rectified current. If a.c. is used, the state of magnetisation at current cutoff might be anywhere around the hysteresis loop. Nevertheless, so-called 'flash transformers' which apply a low voltage, high current (5 V, 500 A) to the specimen through prods (see Section 5.1.3), have been used extensively for residual magnetisation. Their proponents have claimed that by using a very-slow-break, arcing switch, the specimen is always left with adequate residual magnetism. With the magnetising current applied for only 1–2 seconds, only a very lightweight, portable step-down transformer is needed, which is a great convenience for site inspection. Certainly many cracks have been found by this technique and it has been in extensive use for over forty years, but its reliability has been queried by some experts, and its use, when other magnetising techniques are possible, is therefore open to question.

Related techniques which are still used are the condenser-discharge method and the 'flash loop'. The former uses a condenser circuit, discharging in about 0.5 s through a coil or prods and producing a very high discharge current. A long pulse is desirable to allow the induced eddy currents to decay before the conductor current also decays. With short pulses a very large capacitor is necessary.

Current flow techniques

Current from an external source, such as a low-voltage transformer, is passed between two contact points fixed to the surface of a specimen, at a suitable spacing distance. The contact points may be clamps, contact heads on a fixed installation (Fig. 5.2), or hand-held prods (Fig. 5.3). The total current passing

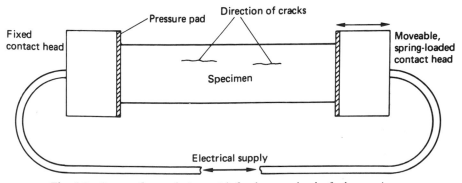

Fig. 5.2 Current flow technique with fixed contact heads, for bar specimens

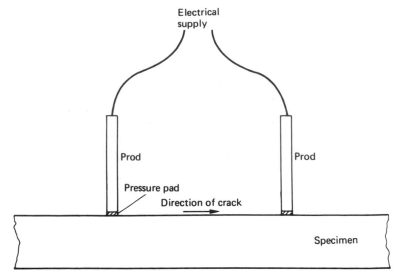

Fig. 5.3   Current flow technique with hand-held prods

through the specimen, usually several hundred or a few thousand ampères, can be measured by meter, but the current density varies from point to point in the specimen, in a complex manner.

For a long, cylindrical bar, with a current, I, flowing along its length between electrodes at the two ends, the magnetic field, H, at the surface is given by

$$H = I/\pi d$$

where d is the bar diameter, so

$$B = \mu_r \mu_0 I/\pi d$$

If $\mu = 200$ and $B = 0.72$ T, for a 30 mm diameter bar, a current of 360 A (minimum) is required.

For many constructional steels in the normalised condition (see, for example, BS:4360: Part 50D), the average permeability can be as high as 700, so that to obtain $B = 0.72$ T (7200 gauss), a current as high as 2400 A m^{-1} (30 oersteds) is unnecessarily large. For non-round bars, an empirical formula, based on the perimeter length of the cross-section (2.4 A per mm of perimeter), is usually used.

Prod magnetisation

The use of hand-held prods to magnetise a local area of a large specimen is a widely applied technique. The magnetic flux distribution is difficult to predict, but can be analysed theoretically by representing the prods (and cables) as two parallel line conductors at right angles to the surface. The surface can be regarded as the electrical equivalent of a mirror and the prods as infinitely long conductors passing through the surface.

Then the magnetic field components at a point P (Fig. 5.4) are:

$$H_{1x} = I\cos\theta_1/2\pi r_1$$

and

$$H_{2x} = I\cos\theta_2/2\pi r_2$$

where I is the current carried in each prod, r_1 is the distance between A and P, and r_2 is the distance between B and P. Summing

$$H_x = H_{1x} + H_{2x} = \frac{I}{2\pi}\left(\frac{\cos\theta_1}{r_1} + \frac{\cos\theta_2}{r_2}\right)$$

This can be plotted as a contour map, but for magnetic testing can be simplified to define a circle between the prods, with APB as a right-angled triangle where

$$\frac{\cos\theta_1}{r_1} = \frac{\cos\theta_2}{r_2} = \frac{1}{D}$$

and D is the distance between A and B. This reduces H to

$$H = I/\pi D$$

Inside the circle, the magnetic field is always greater than the chosen value. Thus, to achieve 2400 A m^{-1}, $I/D = 2400\pi$, which is equal to 7536 A per metre of prod spacing. If the area between the prods to be inspected is less wide, say an ellipse, a much smaller current is necessary, and for straight butt–weld inspection, an empirical rule of 1000 A (rms current) per 250 mm of prod spacing is widely used.

  This theory is simplified and only applicable to flat or near-flat surfaces. The current flowing in the prod heads and the length of the prods have an effect on

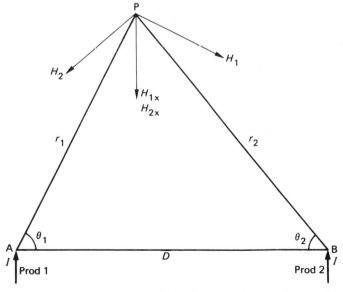

Fig. 5.4   Diagram of resultant magnetic field with two-prod magnetisation of spacing D

the field strength. The orientation of the leads in relation to the line between the contact points is also important. The (prod length)/(prod spacing) ratio should be at least 1:1, and the leads should be at right angles to the line between the contact points.

The equations used so far are valid for all waveforms, but the numerical values are only correct for d.c. For other waveforms they represent peak values and must be converted to meter values. For example, with an a.c. current, soft iron meters and most electronic meters measure rms values, which are $2^{-1/2} = 0.71$ times the peak values. Similar corrections are required for half-wave and full-wave rectified circuitry, both single-phase and three-phase.

With current flow, the magnetic field is at right angles to the direction of the current and as flaws are best detected when they cross the magnetic lines of force, the preferred direction for the current flow (the line between the prods, clamps etc.) is along the expected line of a crack. Normally, the magnetising operation is carried out in several different directions, with a minimum of two directions at right angles. The layout of successive prod positions and spacings has to be determined for individual applications.

Threading bar or cable

A current is passed through a bar (Fig. 5.5) or a cable (Fig. 5.6) threaded through an aperture in a specimen. The obvious application is to ring-like specimens or short lengths of tubing. It should be remembered that it is the inner surface of the bore which is magnetised, with a circumferential magnetic field. Flaws on the outer surface of a tube can also be detected if the wall thickness is not too large, but there may be a large difference in the magnetic field on the two surfaces.

Rigid, threaded bars are usually made of copper to reduce heating. If a flexible cable is used, several turns of cable may be passed through the bore (Fig. 5.6) and the current reduced in proportion to the number of turns. The current, I, required can be derived from

$$H = I/2\pi RN$$

where R is the radius of the inspected surface and N is the number of turns.

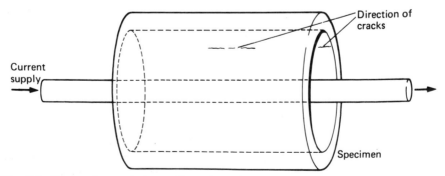

Fig. 5.5   Current flow technique with conducting bar, for tubular specimens and longitudinal cracks

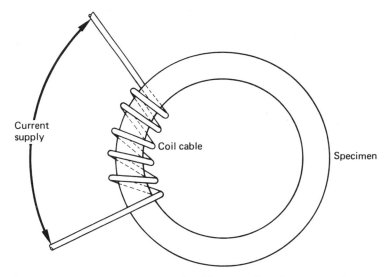

Fig. 5.6 Current flow technique with cable wrapped on ring-type specimens

If H must exceed 2400 A m^{-1} and a threaded bar is used ($N = 1$), a 40 mm inner radius tube will need 600 A.

If the threaded bar is off-centre, a higher current may be necessary, and it is desirable to use as large a diameter of bar as is conveniently practical, within the need to apply the magnetic ink or powder and view the indications.

Rigid coil technique

A specimen, usually bar-shaped, is placed inside a rigid, helical coil (solenoid) carrying a current (Fig. 5.7). This magnetises the specimen parallel to the axis of the coil, so detecting linear flaws at right angles to the axis of the specimen and coil (e.g. cracks running around the curved surface of a bar specimen). A modification of the rigid coil technique is to use a hinged coil which can be opened along its length and used on components such as long pipes which have no accessible end. An analogous technique is to wrap a flexible coil round a bar or tube, again to detect transverse defects on the curved surface.

The shape of the specimen and its position within the coil affect the magnetisation produced. If the specimen radius is r, its length, L, and the coil radius is R, and assuming $r/R < 0.3$ and L/R is large, an empirical formula can be derived for the required current, I, in a coil of N turns:

$$I = 64\,000R/LN$$

for a flux density of 0.72 T and d.c. equipment.

With specimens having a small L/D ratio, there can be a demagnetising effect, and extension pieces on the ends of the specimen may be necessary.

If a coil is wound round the outside of a specimen, the field will be a minimum at a point in the specimen midway between two turns of the coil—point P in Fig. 5.8. At this point, neglecting the effects of all but the two adjacent turns of the coil and assuming the cables are straight and long

$$H = I/\pi(T + Y^2/4T)$$

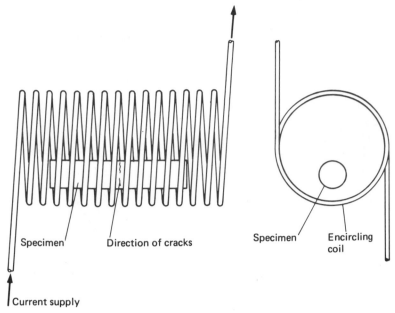

Fig. 5.7　Current flow technique with rigid coil, for bar-type specimens: the specimen may be central inside the coil, or offset as shown

where Y is the spacing between adjacent windings and T is the wall thickness (both in metres). If a.c. current is used, the skin effect limits penetration and BS:6072 suggests replacing T by 10 mm, then putting $H = 2400$ A m^{-1}, and putting T and Y in mm,

$$I = 7.5(10 + Y^2/40)$$

Flexible cable laid on test surface

A relatively new and not yet widely used technique is to have a flexible cable carrying a current, laid on the surface of a large specimen. If the cable is close to the surface, the lines of force are approximately normal to the surface and perpendicular to the direction required for flaw detection, but if the cable is spaced away from the surface by a distance a (say), a strip of width $2a$, immediately under the cable, is effectively magnetised as there is a component of the magnetisation parallel to the surface. The return loop of the cable must be at least $10a$ away (Fig. 5.9). BS:6072 recommends a current in the cable of

$$I = 30a \qquad (a \text{ in mm})$$

This technique has the advantage that there is no danger of burn marks, such as are sometimes caused by prod contact points (arc burns). It is also very convenient for underwater applications of magnetic particle inspection (MPI).

Magnetic flow techniques

Any technique in which the specimen completes the magnetic circuit of a magnet is termed 'magnetic flow'. Thus an electromagnet or permanent

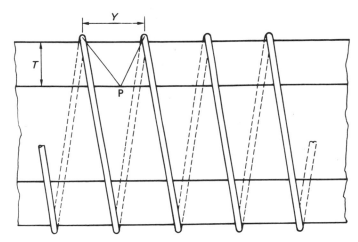

Fig. 5.8 Encircling coil: determination of field strength

magnet placed on the surface of a specimen produces a magnetic field between the pole-pieces of the magnet, which will be cut by a crack lying transverse to the line between the pole-pieces and so a leakage flux is produced (Fig. 5.10). The magnets can be fixed, bench type with solenoid heads, or portable, permanent, or electromagnets. The pole-pieces need to be in good contact with the specimen and flexible pole-pieces are often used. Electromagnets with each pole-piece consisting of a series of rods, or with articulated legs which can

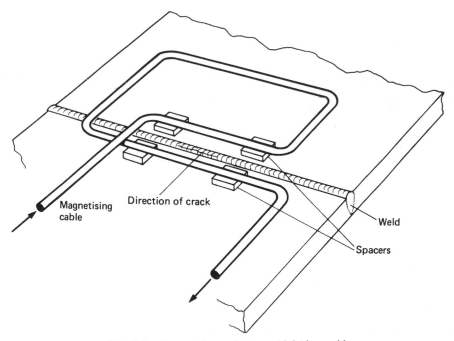

Fig. 5.9 Current flow technique with laid-on cable

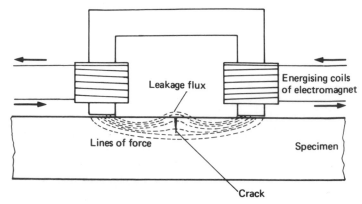

Fig. 5.10   Magnetisation with an electromagnet

move to accommodate the shape of the specimen surface and ensure good magnetic contact, are available.

It is also possible to use the circular magnetic field from a single-leg bar electromagnet placed end-on to the specimen surface. This technique is particularly useful on specimens where access is difficult.

Only the current in the energising coils is under the control of the inspector, so it is usual first to apply a very high current to obtain saturation as evidenced by 'furring', and then work with about one-third of that current. The energising current can be a.c. or d.c. With modern magnetic yokes, fields in excess of 2800 A m^{-1} can be achieved.

With permanent magnets and sometimes also with electromagnets, the level of magnetisation is assessed from the force required to pull the magnet off the specimen, or by the lifting power of the magnet. The pull-off force is the force that has to be applied to one pole-piece to break its adhesion to the surface of the specimen and the lifting power is twice the pull-off force. The pull-off force is directly related to the square of the flux density below the pole-piece and perpendicular to the surface, but this is not the same as the tangential flux density required for MPI.

For electromagnets, BS:6072 specifies a maximum pole spacing of not more than 150 mm and a lifting power of at least 18 kg for magnets to be used for MPI, and in general does not recommend permanent magnets. With a.c. energised electromagnets, a lower lifting power is specified (4.5 kg for a 30 cm pole spacing) and similar values are suggested by the ASTM. The reason for a lower acceptable value is that the skin effect concentrates the magnetic flux close to the specimen surface. Portable electromagnets need a power supply and cables, but are convenient to use, in that they require only one hand, leaving the other free to apply the magnetic ink. The pull of the magnet indicates that good contact has been made with the surface and that the magnet is energised.

Induced current flow techniques

An a.c. current is made to flow in a ring-shaped specimen by effectively making the specimen the secondary winding of a transformer (Fig. 5.11). This secondary winding is a single turn, leading to a very-high, low-voltage

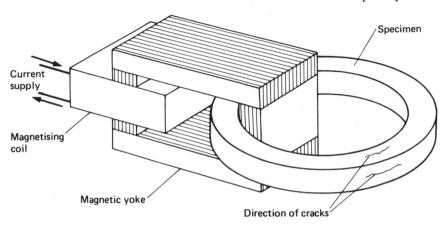

Fig. 5.11   Magnetic yoke magnetisation (induced current flow in a ring-shaped specimen)

magnetising current. The magnetic yoke has to go through the bore of the specimen and this restricts the application of this method to cases where there are a large number of similar specimens, making it worthwhile building special magnetising equipment. It is not a widely-used method, but has the advantage of being a non-contact technique. The magnetising flux achieved is difficult to calculate because of the unknown efficiency of the transformer coupling, and is usually measured on a test piece with a clamp-round meter, which measures the current flow.

5.1.4   Equipment

Much magnetic particle crack detection is carried out with portable equipment which is taken to the specimen on site, using mobile current sources. Transportable equipment in which adjustable contacts can be applied to smaller specimens, or be fitted with energising coils, is built for up to 10 000 A output. Systems are commercially available supplying single-phase a.c. (step-down transformers), half-wave rectified, full-wave rectified, and three-phase.

A very strong case can be made for the use of half-wave rectified a.c. magnetisation because of the 'skin effect'. This concentrates the magnetic field on or near to the specimen surface, where it is needed. Also, on irregular-shaped specimens, the magnetic field will follow the surface, whereas a d.c.-produced field may not do so. The only basic advantage of d.c. is the ability to show subsurface defects. Combined a.c. and d.c. magnetisation (Fig. 5.12) will produce a swinging field; if two phase-shifted a.c. fields are combined, a full rotation of the field can be obtained (Fig. 5.13). One type of magnetisation system, using combined a.c. and d.c. fields, applied to a pipe, is shown in Fig. 5.14. With a rotating field system, only one magnetisation is necessary to detect cracks in any orientation.

There are also multi-directional current flow magnetisers in which there are three independent circuits, each capable of supplying, for example, 1000 A. Automatic switching between pairs of prods produces a sequential application of current for a few seconds in several different directions, so enabling surface flaws in any orientation to be detected, at one application of the magnetising

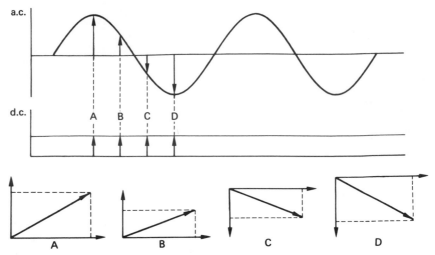

Fig. 5.12   Combination of a.c. and d.c. magnetisation to produce a swinging magnetic field

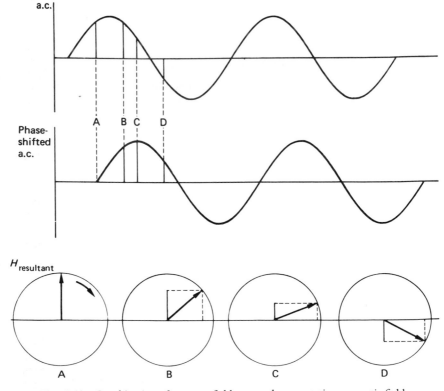

Fig. 5.13   Combination of two a.c. fields to produce a rotating magnetic field

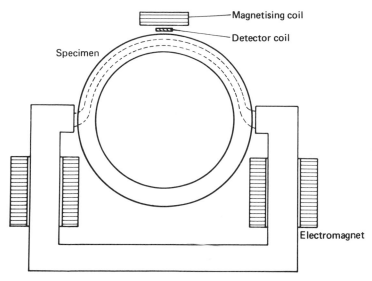

Fig. 5.14   Combined a.c. and d.c. magnetisation equipment

current. Alternatively, using a three–phase supply and three sets of current prods or clamps, a rotating magnetic field can be produced, which also will detect flaws in any orientation. These multi-field systems can greatly increase the speed of magnetic particle testing.

With these combined or rotating field methods, magnetic ink rather than powder is used and it must be applied during magnetisation. It must also be applied as a gentle flow, because a high-pressure application can flush away a crack indication which has been formed only when the magnetic flux is in the optimum direction in relation to the crack. Experience has shown that these combined field methods can detect cracks which were missed by two separate applications of the magnetising field.

The current flow method with prods or clamps is one of the most-widely-used methods of magnetisation. If the prods are hand held there is always a danger of arcing or burning at the contact points. The test piece must be clean to ensure good electrical contact and the prod points are usually fitted with copper mesh pads to assist electrical contact. Lead contact covers on the prods are also useful, but local overheating of the lead can produce noxious fumes and lead covers should only be used in a well-ventilated environment. Contact pads should be cleaned regularly and replaced when damaged. Copper and lead burns can produce metallurgical problems on certain alloys.

On fixed-head machines, the contact heads are usually pressed on to the specimen hydraulically, and there is less of an arcing problem, but copper gauze or mesh contact prods are still used rather than solid-metal contacts.

5.1.5   Flux indicators

It is not possible to measure the magnetic flux in the surface of a specimen, and the use of flux probes for MPI has not become widespread. There are two types of flux indicator which are used.

(1)   The Burmah-Castrol flux indicator—this consists of a sheet of 0.10 mm carbon-free iron, in which three knife-slits have been cut, sandwiched between two sheets of 0.05 mm brass (Fig. 5.15(a)). This is laid on the specimen surface with the slits oriented across the magnetic lines of force, and sprayed with ink or powder during or after magnetisation. The three slit lines should be clearly delineated.

(2)   The Berthold segment-type flux indicator (Fig. 5.15(b))—in this, the non–magnetic gap is spaced away from the surface.

Hall–effect meters containing a small crystal of a semiconductor can be calibrated and used as field strength meters, but are not very convenient for field use.

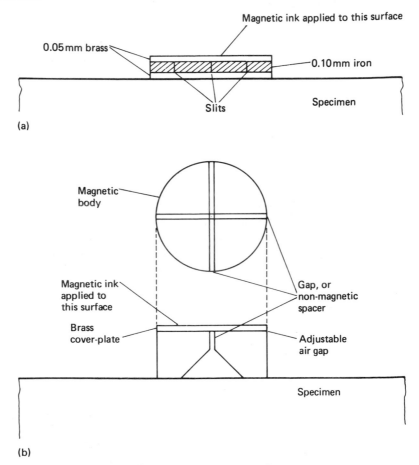

Fig. 5.15   Magnetic flux indicators

5.1.6   Magnetic inks and powders

After the specimen, or a local area of the specimen, has been magnetised, the magnetic ink or powder, containing finely-divided ferromagnetic particles, is applied. The magnetic particles have a low magnetic remanence, so that they do not unduly stick together. They also have a high permeability so that they produce an indication with small leakage fields. They must be applied liberally, so that they can move over the magnetised area and collect on the regions where there is a flux leakage.

In the past, magnetic powder was more widely used than magnetic ink in the USA, and the opposite was true of Europe. Today, magnetic inks appear to have the ascendancy over powders. Given good-quality materials, there is little to choose between inks and powders, but with inks it is possible to formulate a material which takes a long time to 'settle' on the line of a crack, or an alternative material which shows the crack much more quickly, but is not so sensitive to very fine detail.

Magnetic ink is a suspension of ferromagnetic particles in a carrier liquid. The particles are usually black iron oxide, with a low retentivity and a concentration of about 2% by volume in non-fluorescent inks. A high-quality ink or powder must compromise on several related properties—particle size, size distribution, particle shape, retentivity of the particle material, and particle mobility. Particle size ranges from about 1–25 μm, with a mean size of around 6 μm for a high-quality ink; the smallest particles tend to aggregate. The particles should contain a mixture of roughly spherical and columnar material, and although many standards still specify settlement procedures to ensure that the particle content of an ink is satisfactory, inks can be formulated which do not settle out, except over very long periods.

The magnetic particles in both inks and powders can be coloured to make indications more clearly visible against the background, and fluorescent inks and powders are also available. The fluorescent pigment must be bonded to the magnetic particle and particle sizes tend to be larger than in black magnetic inks. With fluorescent particles the viewing of the specimen is done in ultraviolet light in the UV(A) range (see Chapter 7). Some inks are both coloured and fluorescent, so that indications can be seen in daylight as well as in fluorescent light and some inks fluoresce in most artificial lighting conditions except sodium lighting.

Magnetic powders for dry application tend to have larger particles, up to 150 μm diameter, and are not so suitable for the detection of very fine cracks.

The carrier fluid is usually a light oil, such as Kerosene, or it can be water containing a wetting agent and a corrosion inhibitor. There are now national standards in some countries for corrosion inhibitors, and much of the prejudice against water-based inks is unfounded.

Basically, magnetic particles are dark red or black, hence the commonly used term 'magnetic ink'; often a metal surface is given a white coat of 'contrast paint', so that a high contrast is obtained between the indication and the background. This contrast paint can be any white, quick-drying paint: if the inspection area is adequately ventilated, a paint of zinc or titanium oxide powder with a binder, in acetone as a carrier, is very satisfactory and dries in less than 10 s, but on some steels, zinc is not acceptable. Contrast paint is best applied with an aerosol, as this results in a thin, even layer, which does not accumulate in grooves, weld undercut, etc. For good results, the paint thickness should not exceed 10 μm. Coatings thicker than this can seriously affect crack sensitivity. Coated, coloured magnetic particles are sometimes claimed to be less satisfactory than black ink or powder, as the particle size may be slightly larger, but this effect is offset by the improvement in crack indication visibility.

Magnetic inks are applied by spraying, hosing, or dipping. It is important to apply the ink liberally and to allow it to flow slowly across the area of surface which has been magnetised, so that the magnetic particles can migrate to the flaws. The flow of ink must cease before magnetisation ends, so as to prevent the indications being washed away. High-power sprays and over-powerful jets are not suitable. Unless residual magnetism is being used, the ink is applied while the magnetising field is present. It is common practice to apply a wetting coat of ink, apply the magnetising field, and then apply more ink. Some technical experience is needed in applying the ink, but more errors are made by not using enough ink, or using ink with inadequate particle content, than from any other cause.

Magnetic inks settle out in storage and very thorough agitation is needed for remixing to produce a good suspension; they are also widely used in aerosols. Inks can be degraded by contaminants off the specimen surface; for example, water in an oil-based ink leads to flocculation, and oil contamination can cause a loss of performance in fluorescent magnetic inks.

Dry magnetic powders should be free-flowing and the maximum particle size is usually specified as in the range 5–100 μm. The particles should have a range of sizes and shapes. Coated particles, providing a range of colours, are available, as are fluorescent dry powders.

Special blowers, with a very low velocity, are used as powder applicators, so that the powder floats on to the surface. The specimen must be magnetised while the powder is airborne, as it loses much of its mobility when it has settled on the surface. It is advantageous to blow a light current of air over the powdered surface to remove excess powder, without disturbing any flaw indications which have formed.

The relative merits of inks and powders depend more on the quality of the individual product, and skill in application, than on any intrinsic merit of either material. Powders are often used on rough surfaces, such as on castings and undressed welds, and can be used on hot surfaces up to 300°C. Inks are usually preferred on good surfaces, on oily or wet surfaces, and for underwater applications.

5.1.7   Surface finish

Dirt and scale should be removed and, preferably, grease and oil, also. It is possible to use magnetic particle inspection through a thin layer of plating or paint without much loss of flaw sensitivity. If very small, fine cracks are to be detected, a good, clean surface is necessary. Surface roughness tends to cause some stray surface leakage flux, which produces a background indication tending to mask real flaws: the effect is analogous to noise in an optical image. On such noisy backgrounds, the variability between inspectors increases, but it is difficult to justify grinding and polishing over large areas. Local grinding on a questionable indication is, however, an excellent way of checking.

Many specifications recommend that the surface roughness should lie within the range 1.6–6.4 μm(R_a) for machined parts, for high-sensitivity magnetic particle crack detection.

5.1.8 Viewing

After magnetisation and application of the magnetic ink or powder, the area has to be carefully examined. Until very recently, this was always done directly, by eye. Obviously, the eyesight of the inspector, lighting conditions, and the accessibility of the surface must be satisfactory. Also, the workload must not be too high to prevent the inspector from having a reasonable viewing-time.

The level of illumination should be 500–1000 lux as a minimum at the surface (this is equivalent to an 80-watt fluorescent tube at one-metre distance). The angle of the light and the prevention of glare are important. Fluorescent inks and powders must be viewed in ultraviolet light (UV(A)), sometimes called 'black light'. This UV light is normally obtained from high-pressure mercury vapour lamps with a suitable filter. The room lighting should be reduced to around 10 lux; UV(A) is the wavelength range between 320 and 400 nm and is harmless to skin and eyes.

Magnetic indications can be sharp or blurred, broad or fine, depending on the width and depth of the flaw. No information is normally obtainable from the surface indication of the flaw height or its width on the surface: the magnetic particles clump to produce an indication much wider than the true opening width of the flaw.

The purpose of most magnetic particle flaw detection is to detect surface-breaking cracks, but surface-breaking laminations, laps, flakes, and shuts can also be detected. Non-metallic inclusions in ferritic steel have a low permeability and show up if they are surface breaking, as do some alloy segregations. Subsurface cracks are rarely detected unless the crack tip is within a few millimetres of the surface, and, even then, inspection conditions have to be perfect, as the indication at best is broad and diffuse. Under laboratory conditions, with a polished metal surface, a very high magnetising current, and using a magnetic powder, weld defects such as incomplete root penetration at 20 mm depth have been detected, but the vast majority of magnetic inspections are for surface cracks only.

False indications can occur. If the magnetisation is too high, 'furring' (build-up of magnetic particles) can occur at edges and notches. Magnetic ink can settle in shallow grooves, or scratches, and give the appearance of a crack. If the specimen is rubbed with a sharp corner of another magnetic material, a track can be formed which attracts the magnetic ink or powder. This is known as 'magnetic writing' and is usually the result of accidental abrasion. The heat-affected zone of a weld, or local segregate areas produced by cold working on a casting, can produce changes in permeability which attract magnetic ink or powder and produce a pseudo-indication if the magnetising field is high.

With the advent of small closed-circuit television cameras and digital image processing, several attempts have been made to automate the magnetic particle inspection system, by enhancing the received image digitally and using pattern-recognition systems to find and identify crack-like images.

5.1.9 Sensitivity

Unlike most NDT methods MPI gives little evidence of malfunction, nor is there any recognised method of measuring the performance of equipment or procedures in terms of a minimum detectable crack size. It has been suggested that the minimum detectable crack size is about $1 \times 10 \times 100\,\mu m$ (width $\times$ depth $\times$ length).

In the USA a ring standard (Fig. 5.16) has been proposed for equipment verification. The drilled holes are 1.78 mm diameter and are at different depths below the cylindrical surface; the standard is made of AISI-01 tool steel, from annealed round stock. Using full-wave rectified a.c. various numbers of holes can be shown (Table 5.1).

Table 5.1   Ring standard

Magnetic particle type	Current (amps.)	Minimum number of holes seen
black ink	1400	3
black ink	3400	6
fluorescent ink	1400	3
fluorescent ink	3400	6
dry powder	1400	4
dry powder	3400	7

A recent evaluation of this ring standard has shown inconsistent results. Experimentally, it was shown that on specific cracks, residual magnetism

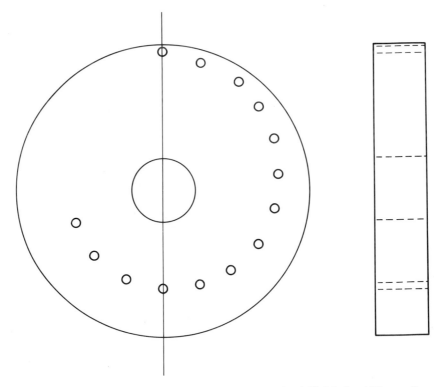

Fig. 5.16   Ring standard. 125 mm diameter, 22 mm thick, drilled holes 1.78 mm diameter. Depths from 1.8–21.4 mm

from 1000 A [1300 A (continuous)] just showed cracks which with 1700 A were easily seen.

Paint and oxide layers greater than 80 μm thick have been shown to cause a large loss in crack sensitivity, although for underwater applications, using an a.c. magnetic yoke, crack detection has been successful through adherent coatings of up to 120 μm thick and paint up to 400 μm.

5.1.10   Image recording

Magnetic indications can be preserved by several methods. Apart from obvious techniques such as television recording and photography, the area under inspection can be covered with a transparent adhesive film which is peeled off, taking with it the magnetic particles. Strippable coatings are also used. Plastic replicas can be made of the surface, including the magnetic indications. Magnetic silicone rubber containing magnetic particles in suspension can be applied to the specimen before magnetisation, left to cure and set after magnetisation, and then removed. This is a particularly useful method on small holes and internal threads.

For photography, particularly with fluorescent inks and powders, there is considerable skill in arranging the lighting and filtration to obtain a high-contrast indication, while at the same time showing some of the background, so that the position of a crack on the specimen can be seen. Matt-white contrast paints provide ideal backgrounds for crack photography.

5.1.11   Automated systems

It is nowadays fairly straightforward to semi-automate or fully automate MPI equipment. In the first case the specimen is positioned manually and the magnetising process then proceeds automatically. Fully automatic systems handling large numbers of similar-sized and shaped specimens have also been built. Robotic handling, automatic control of magnetising current, application of ink and re-positioning can all be computer-controlled. The use of silicon thyristors in power packs instead of silicon diode rectifiers has greatly simplified automatic control of magnetisation current. The quality of the magnetic particle bath can also be monitored.

The first attempts at automated viewing used visible light scanning, followed by UV light scanners and more recently laser light scanning. These systems worked on the same principle as described in Chapter 2, Fig. 2.4. Alternatively, a television camera, usually with an image intensifier and a digital output, can be used to scan the specimen surface after magnetisation and application of magnetic ink. Signal processing is used to discriminate defects from background noise, spurious indications, edges, etc. Much of the success depends on the computer software and the simplicity or complexity of the surfaces to be examined.

The flying spot scanner has the advantage that the laser produces more intense illumination and the receiver can handle a larger contrast ratio than a television vidicon.

Automated viewing brings advantages of reliability and reproducibility, but the crack sensitivity and inspection speed will depend on the size of the pixel used and the complexity of the computer software. False rejection rates of <1% have been claimed.

The combination of numerical modelling of the magnetic flux in a complex-shaped object, even when a combination of several magnetisation circuits is used, together with computer control of the equipment, will probably produce major advances in magnetic testing of mass-produced specimens.

5.1.12 Demagnetisation

Specimens sometimes need to be demagnetised after testing. If a specimen can be placed in a coil having an a.c. supply, this is straightforward. The specimen is gradually withdrawn along the coil axis to a considerable distance, while the coil is energised. Alternatively, a reducing a.c. field can be used, reducing to zero while the specimen is inside the coil.

Total demagnetisation ($B = 0$) is almost impossible to achieve, but usually small amounts of residual magnetism are acceptable.

5.2   MAGNETOGRAPHY

There are alternative methods of revealing a magnetic flux leakage, apart from the application of magnetic ink or powder.

In one method, the magnetic leakage field due to flaws is recorded on special magnetic tape which is pressed on the specimen surface during magnetisation. The tape is then removed for processing and analysis of the indications. This process is called *magnetography*, and while it has been the subject of extensive literature, it does not seem to have been widely used.

The tape used is much thicker (0.5 mm) than conventional magnetic recording tape, and is consequently less easily damaged. After magnetisation

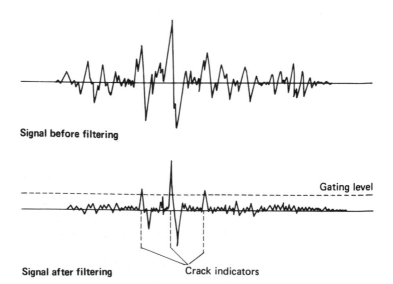

Signal before filtering

Gating level

Signal after filtering   Crack indicators

Fig. 5.17 Magnetography signals before and after filtration

and removal from the specimen, the tape is scanned with rotating cylinders carrying field-sensitive probes and the output is displayed on an oscilloscope screen.

The special merit of this process is that the tape can be 'read' under good laboratory conditions, although the tape itself may have been used underwater, or under difficult site conditions. Further, the cleaning of the weld is not so important, and pseudo-indications and noise can be removed by high-pass filtering of the signals with a cutoff frequency (Fig. 5.17).

5.3   FIELD-SENSITIVE PROBES

The flux leakage field caused by a flaw can also be detected directly by a sensor probe. Two types of probe are widely used: induction coil sensors and solid-state sensors, such as Hall-effect probes or magnetodiodes.

The coil sensor depends on Faraday's law of induction where the induced voltage, V, is given by

$$V = -\frac{\mathrm{d}}{\mathrm{d}t}(NL)$$

where N is the number of turns in the coil, t is time, and L is the leakage flux.

For d.c. magnetisation therefore, the detector coil must be in motion relative to the specimen.

The Hall effect depends on a conductor carrying a current placed in a magnetic field. If the field is at right angles to the current direction, a voltage is generated in the conductor, at right angles to the current and the magnetic field. This voltage is proportional to the magnetic flux. The effect of probe lift-off can be eliminated by using two Hall elements in the probe, one above the other with a fixed, known spacing.

Hall elements are usually bismuth-doped semi-conductors such as indium antimonide. Metals do not make good Hall sensors as their Hall coefficient is relatively low. A Hall sensor, its power supply and an amplifier can be combined in one chip. A Hall-effect probe can therefore be extremely small and also can be used at slow scan speeds.

Magnetodiodes are solid-state elements whose resistance changes with field strength. They consist of p and n zones in a semi-conductor separated by a recombination zone. Magnetodiodes have a flat frequency response up to about 3 kHz and are stable to about 50 °C.

Magnetoresistive sensors have also been devised; these may consist of an evaporated magnetoresistive layer on glass, with attached leads, which is scanned over the magnetised surface. They have a reasonably good sensitivity, but can saturate.

The principal application of flux leakage methods with field-sensitive probes has been to automated high-speed scanning applied to bars, wires and tubes in ferromagnetic material. It is possible to have a rotating magnetisation yoke, rotating inside or outside a tube, with a detector probe between the pole shoes (Fig. 5.18). The probe can have differential output and there is obvious scope for signal processing with electronic circuitry. Calibration specimens with artificial defects have been standardised.

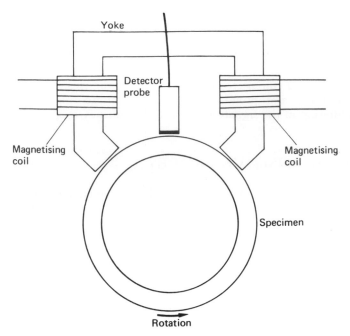

Fig. 5.18   Flux leakage detector on a tubular specimen

A large range of automated flux leakage equipment has been developed by Foerster, using both probe and rotating specimen arrangements (Fig. 5.19). The advantages of such methods are that no coupling medium is necessary, both high sensitivity and high inspection speeds are possible, and defects on the two surfaces of a tube can be separated. Using outside encircling coils the following flaw depths have been claimed:

> 5 mm wall, on outer surface 2% of wall thickness
> on inner surface 5%
> 25 mm wall, on outer surface 2%
> on inner surface 20%.

The disadvantages are that two magnetisation arrangements are needed for longitudinal and transverse defects: there are problems at the ends of the tubes. Compared with automated eddy currents testing (see Chapter 6), magnetic methods do not detect geometric flaws such as dents and wall-thickness changes.

By using a high-frequency a.c. magnetising field, the magnetic flux is concentrated on the surface, due to the skin efffect, and follows the outline of the crack. If the magnetising field is sufficiently high that the surface of the specimen is magnetically saturated, this saturated area follows the outline of the crack and thus effectively increases the 'magnetic width' of the crack and makes it more easily detected. It is claimed that with this technique a linear relationship between crack height and signal can be obtained.

A considerable effort has been put into both direct experiment and theory (finite element modelling) to determine a relationship between crack height

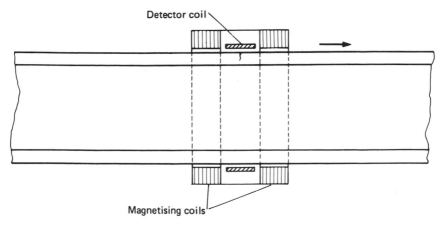

Fig. 5.19   Flux leakage set-up for transverse defects

and d.c. leakage field strength at the surface. The results, at present, are contradictory. Present theories suggest that only under limited conditions is the magnetic leakage field strength at the surface independent of the crack width, but extensive experimental results have shown a reasonably linear relationship with crack height independent of crack width, for a wide range of crack heights, crack widths, and magnetising fields.

Field-sensitive probes are also used with residual magnetism to inspect lengths of oil-field piping (e.g. 7 m lengths) for longitudinal defects.

A very large pulse of current is produced in an aluminium bar along the centreline of the pipe, by a capacitor discharge. Both single-pulse and multipulse systems are used. The circuit is designed to provide near-saturation of the complete volume of the pipe. Theory shows that the operation of this method can be complicated by eddy currents in the pipe induced by the initial rise in the applied current, but by using a long pulse length (about 150 ms) from a high-capacity capacitor, this effect can be minimised. A circuit with an 8 F capacitance, giving a current of 6500 A, together with a series of flux leakage detectors on the outer surface, has been found to be satisfactory. The circuit providing the capacitor discharge includes a silicon-controlled rectifier to suppress oscillations. In the midwall region of the pipe, the field strength is around 3200 A m^{-1}.

The leakage flux signals induced by residual magnetism are small and detection can be complicated by low signal-to-noise ratios, and by random mechanical vibrations of the detector. Low-pass filtering can be used to eliminate the latter, but a better method is to use the signals to modulate a higher-frequency carrier; an a.c. amplifier can then be used, which is tuned to the carrier wave frequency. The sensing coil of the detector is vibrated at the chosen carrier frequency and the modulation is determined by the field gradient in the direction of vibration. However, the spatial resolution of such search coils is limited, and greater resolution can be obtained with an a.c. differential Hall-effect probe, with a tuned-frequency amplifier. Two Hall-effect probes are located about 12 mm apart and supplied with a sinusoidally-varying current (10 Hz): the difference in the two Hall voltages is then measured after the output d.c. is removed by filtering.

In buried pipelines, it is desirable to detect local thinning due to corrosion during the service life, and as this is likely to occur on both inner and outer surfaces, special magnetic testing techniques are necessary. The magnetisation has to be applied from the inside of the pipe, usually with an electromagnet, so that magnetic field strength leakage variations due to external corrosion are likely to be relatively small. The detectors are carried along the bore of the pipe on a 'pig', and consist of a series of flux leakage detectors, such as flux-gate probes, search coils, or Hall-effect probes.

There are possibilities in direct imaging of magnetic leakage fields, either by the eye or with a camera, by magneto-optical techniques, using magnetic garnet films. In these films, which consist of bismuth-substituted iron garnet, the state of magnetisation is switched in regions of weak magnetic anomalies. Bismuth-substituted iron garnet possesses a large specific Faraday rotation, so that plane-polarised light transmitted through the film has its plane of polarisation rotated by an angle proportional to the local, time-dependent magnetisation.

5.4   MEASUREMENT OF METAL PROPERTIES

Microstructural flaws are responsible for mechanical properties and the magnetisation behaviour is also a function of these flaws in ferromagnetic materials. Thus magnetic properties can be used to measure or monitor the properties of some steels. This technique is particularly widespread in the USSR. Some magnetic parameters, such as Curie temperature and saturation magnetisation, are relatively insensitive to changes in microstructure: these properties will change with chemical and phase composition. Properties related to magnetisation—e.g. permeability, coercive force, and Barkhausen noise—are more strongly dependent on flaws, which also affect mechanical properties.

The chief problem is to build up a correlation between magnetic and mechanical properties for a particular material, e.g. for cast-iron used in vehicle brake discs a linearly proportional relationship was found between Brinell hardness and the product

(permeability × resistivity)

A second problem is the elimination of the previous magnetic history of the specimen.

A coercivity meter can be used in some circumstances to monitor the quality of heat-treatment and some mechanical properties. It is obviously necessary, however, that there is an unambiguous correlation between the tested properties and the coercive force, and this is not always the case (Fig. 5.20). The coercivity meter measures the local coercive force in a given area by using a laid-on electromagnet and a flux-gate ferroprobe. In constructional and plain carbon steels with C > 0.3%, there can be a dip in the curve (curve B in Fig. 5.20), compared with the curve for low carbon steel (curve A).

Another magnetisation parameter, which varies unambiguously with temperature, is the relaxation coercive force, defined as the field intensity required to bring the magnetic material from a state of remanent magnetisation to a statically demagnetised state. The relaxation coercive force decreases

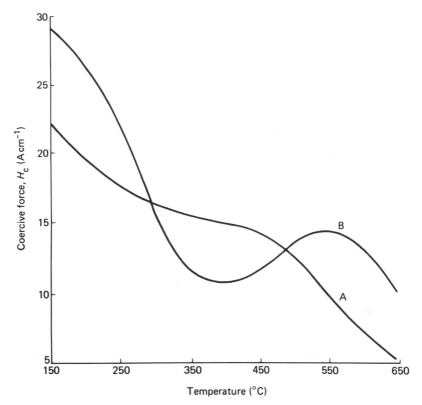

Fig. 5.20   Coercive force plotted against temperature, for two steels: A, 0.18% carbon and B, 0.60% carbon

considerably over the temperature range 150–600 °C, whereas the coercive force shows only a small change over this temperature range. Measurement of relaxation coercive force, however, presents some problems and it may be more convenient to measure remanent induction. This is the induction that results after the demagnetising field is switched off, the specimen being at a point on the descending part of the hysteresis loop. Again, however, with steels having C>0.3%, there can be a peak on the curve with a too-small demagnetising field. Remanent induction can be measured with either a ballistic galvanometer or a flux-gate ferroprobe magnetometer to measure the leakage field.

The relaxation coercive force or remanent induction can also be related to hardness, and can be measured on local areas of large specimens with a laid-on electromagnet with a moving-coil element for sensing the leakage flux. After magnetisation and partial demagnetisation, a current is passed through the coil to produce a constant deflection. The reciprocal of this current is a measure of the remanent induction. The relationship between remanent induction and hardness is not linear, so that calibration curves must be obtained.

The limiting problem in all these uses of magnetic methods to measure or monitor metallurgical properties, is the large number of interacting variables which are involved. Some of these difficulties can be resolved by signal processing techniques.

Grain boundary structure (precipitates, etc.) determines many of the mechanical properties of a metal, and it is these same grain boundaries that affect the magnetic domain structure. This interaction is not yet fully understood, but it seems likely that measurements of magnetic Barkhausen noise and Barkhausen jumps might provide more information on mechanical properties.

Barkhausen noise is due to the movement of magnetic domains in a mass of ferromagnetic material. Magnetisation proceeds as a series of small jumps and if the magnetisation is produced by a coil, and this is connected to a speaker, the jumps can be heard as a series of clicks, or a hissing noise. An analogous effect is 'magnetoacoustic emission', which is due to the transient elastic waves arising from domain wall movements due to magnetostrictive strain. This is measured with a detector on the specimen surface and has the characteristic of coming from the bulk of the specimen, whereas Barkhausen noise is a surface-only effect, due to eddy current shielding.

Both effects have been shown to be sensitive to stress, microstructure, heat–treatment, cracks, and plastic strain, but the relationships are exceedingly complex. In one particular steel (4360 C–Mn), when homogenised and tempered, the Barkhausen noise showed a peak, and the magnetoacoustic emission a double peak with tempering temperature. In fully-annealed material, neither effect showed a peak; in normalised material, only the magnetoacoustic emission showed a peak. It would seem that pearlite, cementite, ferrite, and martensite all have different characteristics in relation to both Barkhausen noise and magnetoacoustic emission, hence the extremely complicated relationships. Barkhausen noise measurements have been used to measure case-hardening depths up to 2 mm on induction hardened surfaces.

Residual stress causes anisotropy in the magnetic permeability and thus in the induced voltages in a coil over the surface of the specimen when the specimen is magnetised. By having two measuring coils at right-angles, either on a tripod-shaped core, or by coil rotation, the induced voltages can be measured and compared. Stress measurements of the order of 50 ± 5 MPa have been claimed.

5.4.1   Harmonic analysis

If a sinusoidally varying field is applied to a ferromagnetic specimen, the induction waveform is not sinusoidal, but has a complex shape which may be Fourier analysed and represented by a number of harmonics, whose phase and amplitude characteristics are related to the magnetic properties of the specimen. Hence the B–H loop may be written as

$$H = H_{\max} \sin \alpha$$

$$B = B_{\max}[A_1 \sin(\alpha + \phi_1) + A_2 \sin(2\alpha + \phi_2) + A_3 \sin(3\alpha + \phi_3) + \dots]$$

where $\alpha = \omega t$, A_1, A_2, A_3, ... are harmonic peak values, and ϕ_1, ϕ_2, ϕ_3, ... are phase angles with respect to the applied field.

If this harmonic structure is measured continuously in terms of amplitude and phase, it can be used to monitor the metallurgical state of the specimen.

This technique has been applied to the cooling of a rolled bar after hot working.

5.5   FERROGRAPHY

As moving parts of machinery wear, some metal is carried away from any contacting surfaces by the lubricant. As wear gets more severe, the wear debris increases and particle sizes get larger. *Ferrography* is a technique which is used to remove this debris from the lubricant and analyse the quantity, size, and shape of the particles to determine whether they are due to normal wear, corrosion, contamination, or lubrication faults, etc.

Ferrography separates out particles having a positive magnetic susceptibility by applying a strong magnetic field to a sample of the lubricating oil running across a glass substrate. The largest particles are precipitated first and the others later, according to size and magnetic susceptibility. The flow of oil is at right angles to the magnetic field, so arranging the debris into rows of particles for microscopic examination. The size and number of particles can be monitored automatically to provide a 'wear particle concentration number'.

Further reading

BS:6072:1981 (1986), *Method for magnetic particle flaw detection*, British Standards Institution, London

Davis N, Magnetic Analysis Techniques, in *Research Techniques in NDT*, Volume 2, Sharpe R S (ed), Academic Press, 1973

Forshaw M E and Mudge P J, Optimization of MPI, *Proc 4th Europ Conf NDT (London)*, **2**, 869, Pergamon Press, Oxford, 1988

Hagemaier D J, *Magnetic Particle Ring Standard Evaluation*, Douglas Aircraft C., Long Beach, USA, Report LR 11711, 1986

Langman R, Measurement of stress by a magnetic method, *Proc 4th Europ Conf NDT (London)*, **3**, 1783, Pergamon Press, 1988

Non-destructive Testing Handbook (2E), Volume 6, *Magnetic Particle Testing*, McIntire P (ed), American Society NDT, Columbus, Ohio, 1989

Non-destructive Testing Handbook (2E), Volume 6, *Electromagnetic Field Modelling*, list of references, 136–145, American Society of NDT, Columbus, Ohio, 1989

Mikeev M N and Gorkunov E S, A review of the physical basis of magnetic structure analysis, *Soviet J of NDT*, **17(8)**, 579, 1978

PD:6513:1985, *Magnetic Particle Flaw Detection*, British Standards Institution, London

Stadhaus M *et al*, System performance control in MPI, *Proc 4th Europ Conf NDT (London)*, **2**, 869, Pergamon Press, 1988

Tiito K and Karvonen I, Evaluating heat-treatment defects, grinding burns and stresses by the Barkhausen noise method, *Proc 4th Europ Conf NDT (London)*, **3**, 1984, Pergamon Press, 1988

Standards on MPI by ANSI, API, ASTM, ASME, AWS, SAE, DoD USDE—all USA, BSI (UK), DIN (W. Germany)

6

Electrical methods

6.1 EDDY CURRENT METHODS (FOUCAULT CURRENTS)

There are a number of electrical methods which can be used for non-destructive testing, such as resistance measurement, electrical conductivity measurement, and the use of triboelectric, thermoelectric, and exoelectron effects, but the two major groups of tests under this general heading are *eddy current testing* and *potential-drop methods of crack detection*.

All electrical methods are indirect: the material properties to be measured have to be correlated with appropriate electromagnetic properties. However, this correlation is often very good, provided that the test conditions are well controlled, but even so, instrument calibration on test specimens is usually essential. The chief limitation of these methods is in finding an electromagnetic property which is a function of one and only one material property.

Eddy current testing involves the observation of the interaction between electromagnetic fields and metals.

The basic requirements are:

(1) a coil or coils carrying an alternating current,
(2) a means of measuring the current, or the voltage, in this coil,
(3) a metal specimen to be tested.

The test coil can be either a single coil, or a pair of coils used as a bridge balancing circuit. The coil can be held in a probe which is moved over the surface of a specimen, or can be wound on a bobbin to move along the inside of a tube or hole, or can be wound in solenoid form to encircle a specimen such as a bar or tube.

An alternating current through the coil, of a chosen frequency, produces eddy currents in the specimen, which modify the exciting current and the resultant current is then related to some of the properties of the specimen.

The size of the resultant current is affected by many variables:

(1) electrical conductivity of the specimen—σ,
(2) stand-off distance between the coil and the specimen,
(3) flaws in the specimen,
(4) magnetic permeability in a ferromagnetic specimen—μ,

(5)   metallurgical variables in the specimen,
(6)   primary coil size (number of turns, diameter),
(7)   a.c. frequency, f,
(8)   dimensions of the specimen,

and one of the major problems in eddy current equipment design is to isolate the significant variables.

An alternating current in the coil produces an alternating magnetic field oriented perpendicular to the direction of the current, parallel to the axis of the coil. If the magnetic field of the coil intersects a specimen of electrically conducting material, eddy currents are induced in the specimen close to the surface, normal to the magnetic field. These eddy currents in turn set up a magnetic field in opposition (i.e. out of phase) to the primary field in the coil, causing a partial reduction in the field of the coil. This decrease in magnetic flux through the coil causes a change in coil impedance, and eddy current testing is essentially the measurement of this change in impedance.

The basic method of testing therefore is to move a probe over the specimen surface, or move a specimen through an encircling coil, or rotate the specimen or the coil. One important practical advantage of eddy current testing is that physical contact between the test coil and the specimen surface is not necessary.

Traditionally, eddy current equipment has used a simple meter display, this being relatively inexpensive and fairly simple to use. The meter reading is sensitive to all the variables listed above and even with skill and experience it is difficult to establish the existence of a crack in the presence of rough surfaces, adjacent fasteners, edges, etc.

The first development was the linear timebase display on a CRT screen. The spot sweeps horizontally with time and changes in the eddy current field cause a vertical deflection on the trace. Using a differential type of probe (see below) there is no deflection so long as both halves of the probe are on sound metal, but as the probe passes over a crack there is, first of all, a downward spike followed immediately by an upward spike. A more advanced form of display, which makes use of phase information as well as voltage change is the vector point (impedance plane) display. The electron beam on the CRT is focussed to a single point in the centre of the screen when the test coil is on sound material. Then the direction in which the spot moves from the balance position is an indication of the parameter causing the change (Fig. 6.1).

6.1.1   Basic principles

The test coil can be characterised by two electrical quantities:

(1)   the ohmic resistance, R,
(2)   the inductance, L.

The reactance is

$$X_L = 2\pi f L$$

where f is the frequency of the a.c. current and L is the self-inductance of the coil. Then, in the absence of capacitance, the resistance and inductance combine together to form the impedance, Z.

$$Z = [R^2 + (2\pi fL)^2]^{1/2}$$

Impedance, Z, is a complex, vector quantity, and it is also convenient to write

$$Z = R + iX \qquad \text{where } i = (-1)^{1/2}$$

assuming the coil capacitance to be negligible.

It is common practice to plot inductance on the y-axis and resistance on the x-axis (Fig. 6.2), so that the coil impedance is represented by a point, P. This plot is usually called the 'complex impedance plane', where R is the real component of the impedance and L is the imaginary component, θ being the phase angle

$$\tan\theta = 2\pi fL/R$$

If a specimen is introduced into the field of the test coil, the original field is modified by the superimposed eddy current field, and this is analogous to

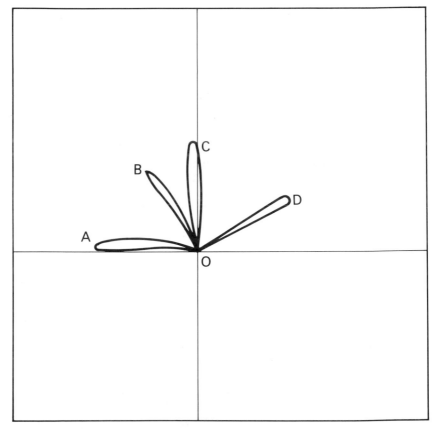

Fig. 6.1   Example of impedance plane display
A – effect of lift off
B – effect of score mark
C – effect of light fatigue crack
D – effect of material composition
O – balance point

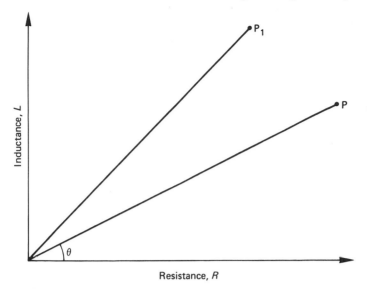

Fig. 6.2   Complex impedance of a coil carrying an alternating current

changing the characteristics of the test coil, so that it has a new impedance Z_1 corresponding to point P_1, with new values of resistance, inductance, and phase angle.

6.2   IMPEDANCE PLANE DIAGRAMS

Historically, the form and magnitude of impedance changes have been studied by means of impedance plane diagrams derived for various geometries. These diagrams are necessarily complex due to the need to incorporate the effects of a large number of parameters. Most impedance plane diagrams are originally due to Foerster (1954), Stanford (1954), Hochschild (1958), Libby (1956, 1971). With the development of computer modelling techniques, such diagrams can now be replaced by numerical methods, either using finite elements or finite differences or methods of moments (Harrington, 1968) and these enable more complex geometries and material changes to be studied (Becker, 1986).

A few examples of experimental impedance plane diagrams, to illustrate their uses, are shown in Figs 6.3–6.7.

If the resistance and inductance of the coil in free space are taken as R_0 and L_0, normalised impedance plane diagrams can be produced by plotting $\omega L/\omega L_0$ against $R/\omega L_0$, where ω is the angular frequency ($=2\pi f$).

An example of part of an experimental impedance plane diagram for a long solenoid encircling a cylindrical bar is shown in Fig. 6.3, to illustrate how impedance diagrams may be used. The numbers along the curve are frequency, f. The small lines represent the directions of changes in impedance due to changes in the bar diameter and the metallurgical condition of the bar (in this example, case-hardening), the inward-pointing lines being metallurgical

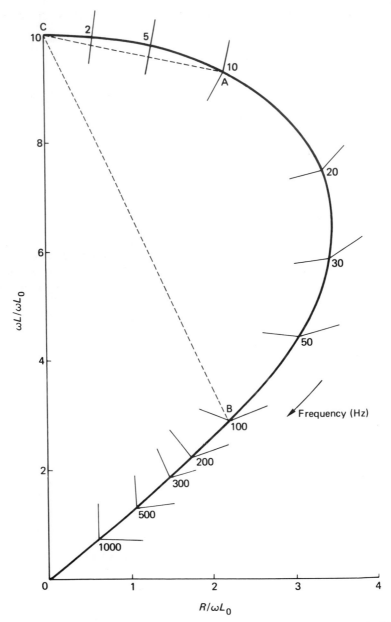

Fig. 6.3 Normalised impedance plane curve for a long solenoid around a cylindrical bar, showing the direction of impedance value changes at different frequencies for small changes in bar diameter and changes in metallurgical structure (case-hardening)

changes and the outer ones a small diameter change. The figure shows that at a frequency of 10 Hz (point A) the effects of diameter change and case-hardening depth would each produce an impedance change approximately along the same line, whereas at a frequency of 100 Hz (point B) the effects of these two variables would be in directions about 135° apart. For simplicity, this curve has been drawn for one value of fill-factor, η, where

$$\eta = \left(\frac{\text{specimen diameter}}{\text{coil inside diameter}}\right)^2 = \left(\frac{D}{D_1}\right)^2$$

In eddy current testing, there are many variables, and if they are all included in impedance plane diagrams, these can become very complicated. Thus it is more convenient to plot the effect of frequency as a generalised term, k, in which

$$k = (f\mu_0\sigma)^{1/2}$$

where σ is the electrical conductivity and μ_0 is the permeability of free space $(=4\pi \times 10^{-7}\,\text{H}\,\text{m}^{-1})$, or to use a ratio f/f_g, as proposed by Foerster, where

$$f_g = 2/(\pi\mu_r\mu_0\sigma D^2)$$

μ_r being the relative permeability ($=1$, for non–ferromagnetic materials).

Simplified impedance diagrams can also be plotted with some of the variables fixed, and Fig. 6.4 shows a normalised impedance plane diagram for a single frequency, for a fixed size of specimen, with different conductivities. The $R = 0$ point corresponds to a coil in air, and as the coil is brought closer to the conductor, the impedance value moves along one of the broken lines, until it reaches a point on the conductivity curve corresponding to the conductivity

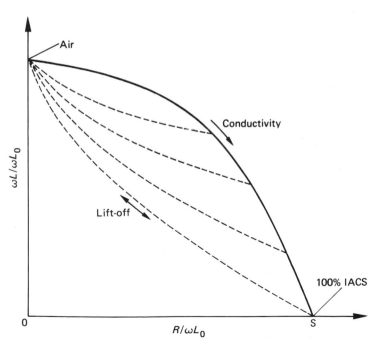

Fig. 6.4   Normalised impedance plane diagram for a coil laid on a plane conductor, specimens of different conductivities, for a single frequency (100 kHz). Point S is for a material of 100% IACS (International Annealed Copper Standard – taken as a perfect conductor)

value of the specimen. The broken curves therefore show the effect of lift–off of the probe.

Figure 6.5 shows a normalised impedance diagram for a coil around a solid non–ferromagnetic cylinder. The broken lines represent equal values of f/f_g and the solid lines are different fill-factors.

For ferromagnetic materials (Fig. 6.6), $\mu_r \neq 1$. The solid lines represent different values of σ for fixed values of μ_r, and the broken lines values of μ_r for given values of σ. The effect of different frequencies on the components of the complex impedance value can be determined from these curves. If it is possible to operate at point A by suitable choice of frequency, etc., the changes in impedance due to small changes in σ (along the curve) will be almost entirely in reactance.

The effects of other parameters, such as the depth of a surface-breaking crack, on a fixed size of specimen, can be added to these curves, and this has been done experimentally by Foerster, using a model consisting of a tube filled with mercury to simulate a conducting specimen, with plastic strip inserts to represent the crack. The impedance diagram curves were then used to determine the conditions at which crack depth could be related to operating frequency independently of changes in the other parameters, D, σ, and η (Fig. 6.7).

If a ferromagnetic specimen is magnetised to saturation, $\mu_r \rightarrow 1$, and curves such as Fig. 6.5 are applicable, but in practice saturation is rarely reached.

Similar families of curves can be produced for other geometries—coil on flat surface, internal and external coils, tubular specimens, etc.

At different points round the impedance curve, corresponding to different frequencies, the effect of, for example, lift-off of a probe from a surface produces a change at different angles. If the output from an individual frequency is 'rotated' and then mixed with that from another frequency, it is possible to suppress the effects of unwanted variables: this is the dual-frequency technique. Similarly, if the impedance display is rotated until the effect of a crack is oriented in the y-direction, the effect of lift-off being in another direction, an amplification can be applied to the y-direction only, thus enhancing this signal at the expense of other, unwanted signals.

With ferromagnetic specimens, the magnetic permeability is not constant at different points around the hysteresis curve and although eddy current inspection in magnetic materials is improved by superimposing a d.c. field on the a.c. measuring field, this d.c. field does not usually produce complete saturation. The relative magnetic permeability to be used in eddy current calculations is therefore the 'reversible permeability', represented by a small secondary hysteresis loop starting from the static magnetic conditions (Fig. 6.8). As seen in Fig. 6.8, the reversible permeability decreases along the initial magnetisation curve, and on the descending part of the hysteresis loop the value peaks at a point where the field equals the coercive field, H_c. The impedance of a coil used on a ferromagnetic specimen depends therefore on the position on the hysteresis loop, according to the applied d.c. field. For calculations, the value of reversible permeability as measured on d.c. should be used. The practical value of reversible permeability depends on both the external magnetising field and the magnetic history of the specimen. If the specimen has an undetermined magnetic condition, a d.c. field having a value of about twice the coercive field, should be applied in one direction and then

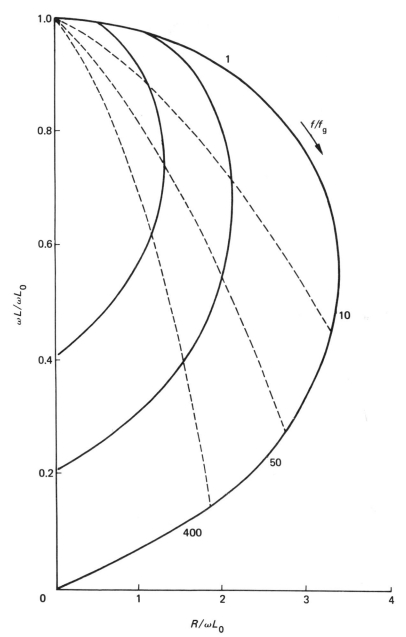

Fig. 6.5 Normalised impedance plane diagram for a coil around a solid non-ferromagnetic cylindrical specimen, for different values of f/f_g and different fill-factors. The broken lines represent equal values of f/f_g

the a.c. measurement made with a similar d.c. field in the opposite direction.

Impedance plane curves can be derived theroetically, starting with Maxwell's equations, but except for the simplest geometry it is a complex calculation, and becomes even more complex if $\mu \neq 1$, or is not constant.

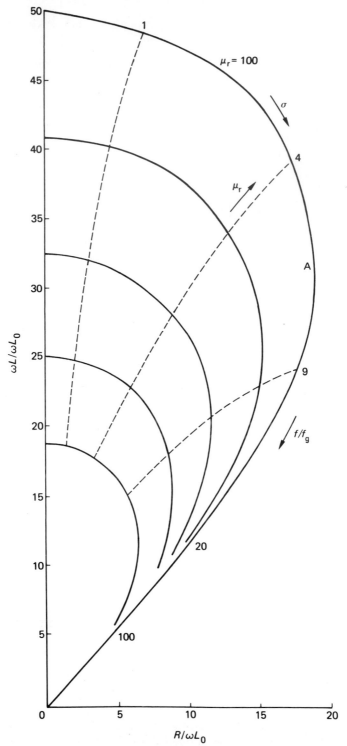

Fig. 6.6 Normalised impedance plane diagram for a coil around a solid ferromagnetic specimen for different values of f/f_g, μ_r and σ

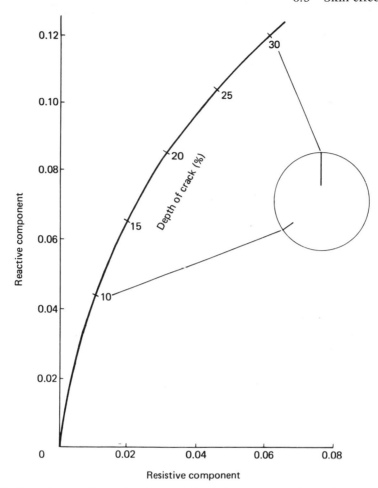

Fig. 6.7 Part of an impedance plane diagram for surface-breaking radial cracks in a cylindrical bar with an encircling coil, where $f/f_g = 15$, $\eta = 1$ and the (crack height)/(crack width) ratio = 25 (after Foerster)

6.3 SKIN EFFECT

When eddy currents are present in a specimen they tend to be concentrated near to the surface adjacent to the coil—the *skin effect*.

The standard depth of penetration, δ, has been defined as the depth at which the intensity of the eddy current is reduced to $(1/e)\%$ of its surface value, where $e = 2.718$, and the value of δ is given by

$$\delta = (\pi f \mu \sigma)^{-1/2}$$

Thus the depth of penetration for a given material is controlled by the frequency, f. Typical values are shown in Table 6.1. The value of δ, in relation to the specimen thickness, is a useful method of determining the frequency to be used in eddy current applications.

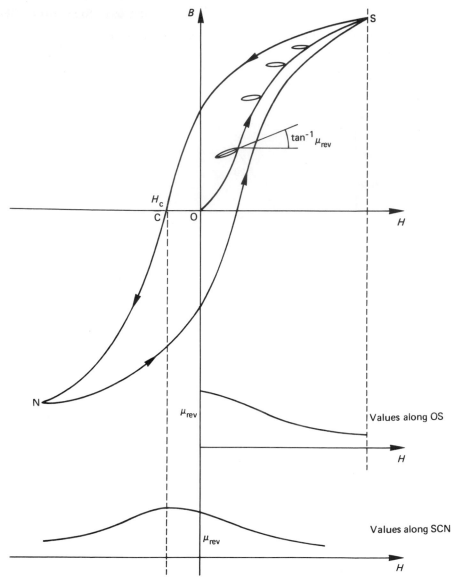

Fig. 6.8 Hysteresis curve, showing the change in the value of reversible permeability, μ_{rev}, at different points along the hysteresis loop

Table 6.1 Depth of penetration, δ (mm)

Material	Frequency (MHz)		
	0.001	0.010	0.050
Copper	2.00	0.64	0.28
Aluminium	2.65	0.84	0.04
70:30 copper:nickel	10.00	3.15	1.41
Titanium	12.00	3.80	1.67
Cast steel	0.50	0.15	0.07
Graphite	45.00	13.00	6.20
Zirconium	12.00	3.50	1.90

6.4   PROBE DESIGN AND INSTRUMENTATION

The design of an eddy current probe depends to a great extent on the specific application. Eddy current instrumentation has to detect small changes in the

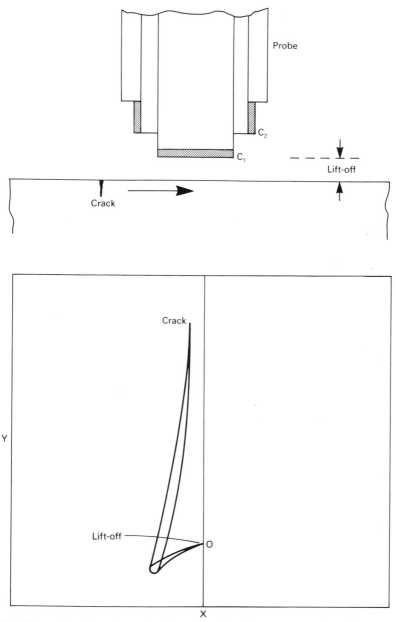

Fig. 6.9   Two–coil probe and X–Y display as a small crack moves across the probe centre–line. C_1 and C_2—probe coils

electrical properties of the coil or coils in the probe. For scanning over surfaces, a simple single coil wound on a ferrite core, or a coil encircling a bar or wire specimen, or a coil wound on a bobbin for insertion into a tubular specimen may be used. These are 'absolute' probes. Lift-off variations of a fraction of a millimetre can produce serious indications capable of masking crack indications, and it is common to design such probes to maintain coil-specimen spacings to within ±0.1 mm.

Double-coil surface probes have been designed by Ghent (1981) to rotate the lift-off signal away from surface defect signals, and it is claimed that these enable surface-breaking cracks down to 0.1 mm deep to be detected. These probes have two co-axial coils connected differentially (Fig. 6.9), with the inner, smaller coil having many more turns. Impedance plane diagrams have been calculated to show how the angle of the lift-off signal varies for each coil with respect to coil diameter, resistivity and frequency. Because the two coils have different sizes, the lift-off signal from each, due to a small change in lift-off distance, can be the same voltage but with different phase angles. The resultant defect signal from the two coils is dominated by the defect signal from the inner, small-diameter coil, and can be set to be at right-angles to the lift-off signal.

It is still desirable to design the probe movement assembly to maintain minimum variation in physical lift-off, but the improved performance of the two-coil surface probe confers additional advantages. Lower frequency systems can be used, giving a greater depth of penetration. In tubing inspection, circumferential as well as axial flaws can be detected. Another form of absolute probe, using a single test coil, has a toroidal reference coil matching the characteristic impedance of the test coil to within 5% over a wide frequency range. Such absolute probes remove the complexity of signal analysis associated with differential probes and can detect gradual defects such as erosion and fretting.

Saturation probes can be used on ferromagnetic tubing. These contain a permanent magnet designed to maximise the saturation field over the test coil—the so-called 'magnetically-biassed eddy current' method. Cecco and Sharp (1986) have shown that 99% saturation is desirable, the relative permeability being 1.15 compared with 1.9 at 95% saturation. At 99% saturation the signals from external calibration holes in a tube show the phase rotation with depth that is obtained on a non-ferrous specimen. For thin-walled tubing (up to about 1 mm) this technique is excellent and can be used for both flaw detection and flaw sizing.

For many applications of eddy currents, differential rather than absolute probes are used, and can be used with meter, linear timebase, or vector-point types of display.

One type of differential surface probe (Fig. 6.10) has a single winding on a split ferrite core. As such a probe is moved across a crack there are two indications 180° apart, and if a probe is being rotated rapidly in a tube or rivet hole, these indications effectively produce a crack indication as shown in Fig. 6.11. A wide crack or mechanical damage, because of the loss of metal at the crack position, produces an indication at a different phase angle.

In a differential probe there are two similar coils, and the impedance changes from each are subtracted, producing zero output when both coils are subject to the same conditions. Such a probe permits a higher signal amplification. Also,

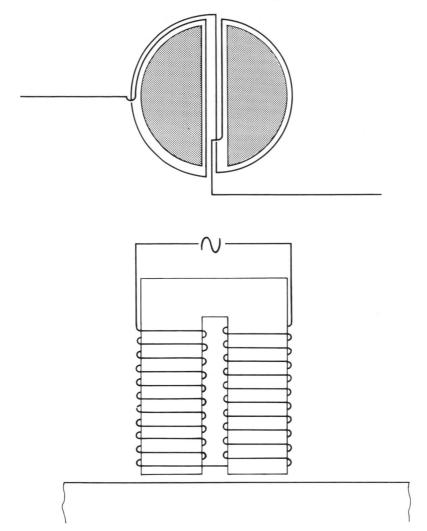

Fig. 6.10   Single split–ferrite–core differential probe

slowly changing variations such as temperature or material properties are automatically cancelled out.

Another probe configuration is to have separate transmit and receive coils, so that the eddy currents generated in the specimen affect their mutual coupling. It is possible to have the two coils differently oriented to detect particular field components, e.g. by having a tangential coil inside a normal exciting coil.

Most of these probe configurations can also be applied to bobbin coils for use inside tubes, as well as to surface probes. A special form of coil for tube inspection, using the 'remote field method' will be described in a later section.

For scanning systems it is possible to use a two-dimensional array of probes, e.g. a 4 × 4 array of ferrite-cored coils. Each coil has its own excitation circuit

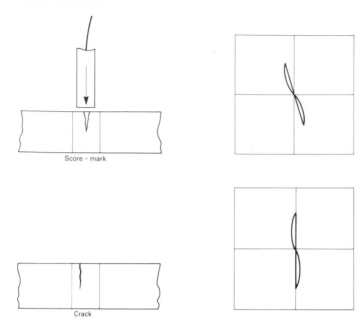

Score - mark

Crack

Fig. 6.11 Rivet-hole inspection with a rotating probe. With a score-mark or a wide crack there is a phase-angle change on the display

with a current-switching technique. At a suitable instant the coil current is deflected through a high impedance resistive load, so producing a voltage spike, the whole being computer-controlled. A liquid crystal graphics display can be used to present the results on which a circle of pixels shows the reading from each probe in the array, so that a crack-like flaw can be seen as a line across the imaged array.

Multiple systems, to examine all surfaces of a complex section such as a rail, can be constructed. By using a combination of fixed and rotating probes, both transverse and longitudinal flaws can be detected. A rotating probe overcomes the problem of flaw orientation.

6.4.1 Instrumentation

The most commonly used form of time-varying coil excitation is a sinusoidal waveform of either single or multiple frequency, and to obtain the maximum information from the received signal, a phase detector is necessary, usually representing the signal as a vector on a complex impedance plane (X–Y) as already illustrated in Fig. 6.1. The vector angle is the phase angle of the impedance.

There are several types of circuit which are phase sensitive. Figure 6.12 shows a simple phase-discriminator circuit.

A voltage V_b, which depends on the measuring coil impedance Z, enters through T_1, usually from a bridge circuit of which the test coil is one arm. Through T_2, a reference voltage V_r is added in the upper half of the circuit and

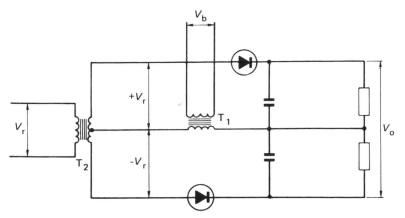

Fig. 6.12   Example of phase-discriminator circuit

subtracted in the lower half; both voltages are then rectified and appear as a d.c. voltage V_o. If V_r has the same frequency as V_b, then

$$V_o \propto V_b \qquad \text{(phase angle } V_r \rightarrow V_b)$$

If V_r is derived from the same oscillator as V_b and is passed through a phase shifter, the output voltage V_o will remain zero whenever the phases of V_b and V_r coincide. It is therefore possible to discriminate against any coil signals of a chosen phase and suppress these.

Waveform choppers and analogue multiplier circuits can also be used as phase discriminating circuits. If the incoming signal is represented as

$$A \cdot \cos(\omega t + \alpha)$$

and is multiplied by phase reference signals $B \cdot \cos \omega t$ and $B \cdot \cos(\omega t + \pi/2)$ the output obtained with the 0° reference is

$$(AB/2) \cdot \cos \alpha + (AB/2)(\cos[2\omega t + \alpha])$$

If then a low pass filter is used to remove the 2ω component, the final X-output is proportional to $A \cdot \cos \alpha$ and the Y-output is derived as $A \cdot \sin \alpha$. A vector rotation circuit with variable-gain amplifiers on the X and Y outputs is necessary to alter the loci of the impedance changes. With multifrequency applications, separate magnitude and rotation adjustments may need to be applied for each frequency component, to obtain the full benefit of multifrequency equipment.

Modern single-frequency eddy current testing uses a wide range of frequencies, from about 60 Hz to several megahertz, but the use of dual-frequency and multifrequency methods has increased rapidly (McNab, 1988). Low frequency methods (400 Hz) can detect a crack in a 1.5 mm aluminium sheet, underneath a 6 mm aluminium covering sheet.

In complex testing situations, where there may be signals from a variety of sources as well as from small defects, additional information is required, and, as some of the signals are frequency-dependant, the responses from different test frequencies can be combined to eliminate unwanted signal components.

In dual-frequency equipment the two excitation signals are combined and applied simultaneously to the test coils of a differential probe, and the differential impedance signal is amplified and then applied to two synchronous phase detectors. By appropriate filtration, four components (phase and magnitude at each frequency) are obtained with a combination of vector rotation and recombination to isolate each component in turn.

With modern equipment the signal is digitised, and microcomputer software can be used to separate the parameters. Signal digitisation permits more accurate calibration and faster signal evaluation, with accurate phase angle determination. Microcomputers can also be used to select a suitable mix of test frequencies to solve a particular problem. Multifrequency equipment is still evolving rapidly, e.g., on ferritic materials, different test frequencies simultaneously applied can mix together, due to the effects of hysteresis, but to overcome this it is now possible to apply the test frequencies separately, and then recombine the results for display.

6.5   PULSED EDDY CURRENTS

The basic technique of pulsed eddy current testing is to apply a short pulse to the transmit coil. The resulting eddy current pulse propagates in the specimen as a heavily attenuated wave of electromagnetic energy with a phase velocity depending on the material and the frequency. For material thickness measurement, a detector coil can be placed on the back surface, or a detector coil can be placed close to the transmit coil with suitable shielding. Usually electromagnetic shielding (ferrite, Mu-metal) is necessary around both the exciting and receiving coils, although this limits the penetrated thickness.

Because the wave propagation is highly dispersive, changes in pulse shape occur with distance into the specimen, and by measuring the time–amplitude characteristics of the received pulse, defect depth can be determined.

Pulsed eddy currents have also been used successfully for the detection of deep flaws (8–10 mm) in austenitic steel. Improved depth sensitivity is obtained at the expense of poor lateral resolution. Most pulsed eddy current equipment converts the received signal to digital data, and as pulse repetition frequencies around 1 kHz are used, a buffer memory is necessary to store data. The received waveform can be digitally filtered, or a discrete Fourier Transform in the frequency domain can be obtained.

6.6   REMOTE FIELD EDDY CURRENT (RFEC) TECHNIQUE

The 'remote-field eddy current' technique, pioneered by Schmidt (1984) and Atherton (1988), has been developed for the inspection of small-diameter non-magnetic metal tubes, using an internal probe. It is claimed to have the advantage over standard techniques in that internal and external defects in the tube wall are detected with equal sensitivity.

The probe (Fig. 6.13) consists of a bobbin-wound transmit-coil, co-axial with the tube, and a ring of pick-up coils separated axially from the transmit-coil by a distance of approximately 1½–3 tube-diameters.

The electromagnetic field at the pick-up coil consists of two components.

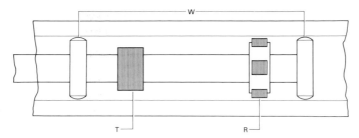

Fig. 6.13   Sketch of Remote Field Eddy Current (RFEC) probe in a pipe.
T – transmit coil
R – array of receiver coils
W– tracking/alignment support wheels

The direct field component, due to the electromagnetic field generated by the transmit coil, remains within the tube: this component (a simple transformer action) decreases exponentially with axial distance from the transmit coil, due to wave-guide attenuation.

Some of the electromagnetic field generated by the transmit coil diffuses through the tube wall near the transmit coil, inducing circumferential eddy currents in this section of the wall: this is the remote field component. The combination of transmit coil current and eddy current generates external dipole fields in a region far away from the tube. The decrease in the external field is more gradual than the exponential decrease in the direct field component, and the external electromagnetic field can be much greater than the internal direct field component. The internal field generated by the diffusion back of the external field through the tube wall, attenuated and phase-shifted, can be detected. As this field has passed twice through the tube wall it provides a through-wall test method with approximately the same sensitivity to external and internal flaws in the tube wall. There is a transition zone between the coils where both fields are significant and in which there are complicated field interactions. In the true remote zone the indirect coupled field dominates and only one vector is involved.

Operation in the remote-zone is normal, but under carefully controlled conditions transition zone operations could have certain advantages such as shorter probes and increased sensitivity.

The use of shields mounted on the probe assembly between the transmitter and detector coils allows reduced separation of the coils, and so can shorten probe length. Laminated disc-shaped aluminium-ferrite rings which do not need to touch the pipe-wall have been proposed (Atherton, Czur & Schmidt, 1989). One application has been to thin-wall zirconium-niobium alloy tubing using a frequency (10 kHz) which gives a skin depth approximately equal to the wall thickness. As with other eddy current systems already described, a complex plane display can be used. The effect of lift-off can be phase-shifted so that the quadrature component of the signal is free from lift-off noise and a linear time-base scan display can also be used.

6.7   SIGNAL PROCESSING

For the interpretation of eddy current data, signal processing using digital techniques as already mentioned, is playing an increasingly important role. The continuous waveform from the detector can be sampled at discrete time intervals, and these sample signals can be digitally filtered or analysed for frequency content. Matched filters can be applied to extract known signals from random noise. Inverse filtering can perform the same task and does not require knowledge of the signal content.

Another method of signal data recording and processing is as follows:

As a defect passes through an eddy current coil, it generates closed loops of the form shown in Figs 6.14 and and 6.15, and the loop is measured on a series of radius vectors, P_1 to P_6, as shown in Fig. 6.16. A predetermined level is set for these to eliminate interference signals. P_1 to P_6 are set arbitrarily at 25%, 50%, and 75% of r_{max} and are recorded digitally together with defect position coordinates and an identification code. The microprocessor within the instrument then carries out the defect evaluation in accordance with the phase position and the signal magnitude.

The problem in all eddy current testing, having set up the parameters, is to separate the wanted from the unwanted signals. Phase analysis can usually separate conductivity from dimensional changes in non-magnetic specimens; in magnetic materials, it can display conductivity changes as shifts of phase and permeability, and dimensional changes as amplitude variations.

Alternatively, the X–Y plane can be divided into sectors and pulses can be generated digitally, proportional to the maximum amplitude of the signal in the significant sector AOB (Fig. 6.17) (Stepinski, 1987). Pulse D4 is outside the sector and D3 is only passing the sector and represents a spurious indication.

If an eddy current probe is scanned in a series of lines across a surface a three-dimensional 'map' of the surface defects is obtained. Digitisation of the signals allows image processing and pattern recognition method to be used; e.g. lift-off suppression, noise suppression, resolution enhancement, etc. Inverse filtering will identify defect types. The impetus for much of this work is the high rate of data acquisition necessary to increase material handling speeds.

6.8   APPLICATIONS

The greatest number of applications are to the high-speed testing of small-diameter, machined tube, bar and wire stock. Sensitivity levels are set by means of reference artificial 'defects' in a test specimen, and specimen handling, data collection, data interpretation, and defect marking are all automated operations.

For thin tubing, both encircling coils and internal rotating probes are used, and a major problem in design is to eliminate probe-to-surface distance variations. The internal probe can have differential coils which only register a flaw when it is under one of the probes and not under the other; if a rotating probe is used, the differential coils can be in the circumferential direction. Many of the practical problems originate from vibration and shock due to the

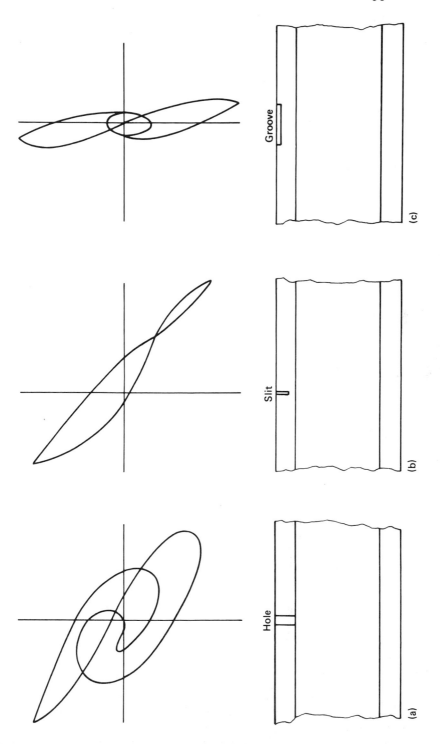

Fig. 6.14 Eddy current signals obtained with an internal rotating probe, operating at 250 kHz: (a) 2 mm diameter radial hole, (b) OD surface narrow slit, and (c) OD surface longitudinal groove

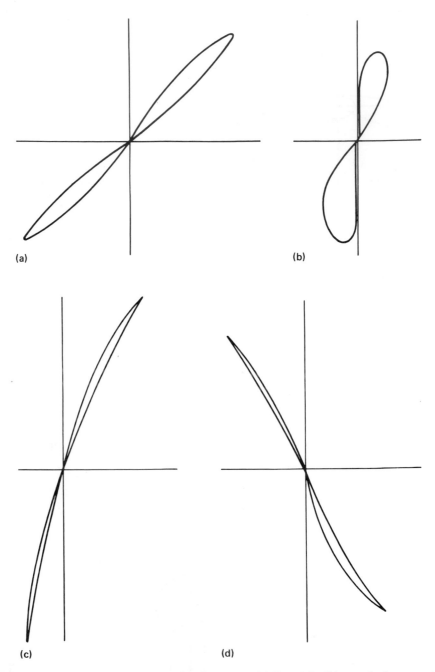

Fig. 6.15   Eddy current signals obtained with an internal differential coil in a steel tube, operating at (100 + 350) kHz: (a) 40% deep OD surface-breaking crack, (b) 0.8 mm diameter drill hole, (c) radial crack, and (d) surface-breaking crack

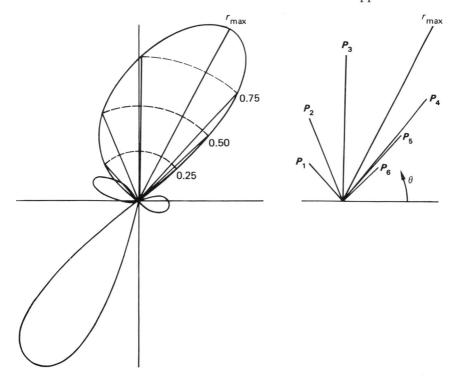

Fig. 6.16 Method of distinguishing a complex eddy current signal: values (r, θ) of **P₁** to **P₆** are recorded from three arbitrary percentage values of r_{max}

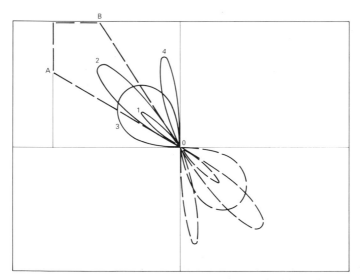

Fig. 6.17 Sector Classification (after Stepinski). Only responses 1, 2, entirely within sector AOB are recorded

tube conveyor equipment, and this can produce pseudo-indications which must be separated from genuine defect indications. A coincidence technique can be used whereby a genuine defect indication is recognised when it occurs successively on two detector coil units spaced a short distance apart. For wire and bar testing, both encircling coil and surface probe techniques are used. The former can be used at very high speeds, up to 70 m s^{-1}. The external surface coil, rotating on a helical scan, is mechanically more complex, and slower, but offers better defect resolution. Differential coils are excellent for detecting short defects, but will not detect long drawing cracks with little variation in depth. Therefore it is necessary to use both differential and absolute measurements, as well as several test frequencies and correlation techniques. Frequently, the wire testing equipment is incorporated into the wire drawing machinery and the testing can take place at high temperatures, with water-cooled probes.

In 6 mm diameter wire, transit coil systems can detect test defects more than about 50 μm (1%) deep, and with a rotating probe, from about 35 μm deep. Defects are normally paint marked for subsequent removal. Eddy current wire testing has been applied to wire as fine as 50 μm diameter (tungsten alloy wire): at a frequency of 120 MHz, and a testing speed of 1 m s^{-1}, cracks 10 μm deep were easily detected.

The more advanced designs of probe and the methods of signal handling and signal processing, using digitisation and computer control as already outlined, have found their way into this type of application. The availability of new computer software packages will encourage further rapid progress in this field of application, particularly in terms of defect identification with so-called 'expert' systems. Real-time digital signal processing is already possible.

At the other end of the applications field, eddy currents are used for the detection of surface-breaking cracks by using a hand-held probe. By having a very small coil in a ball-nosed probe, which is moved over the surface of a specimen, surface-breaking cracks can be detected. As the crack cuts into the eddy current field, the pattern is distorted, so that as the probe is moved across the crack, the instrument produces a trace or instrument reading. Many such simple crack detection eddy current devices have been marketed, and they can have a high sensitivity. A rough guide is that the (minimum) crack depth to surface roughness ratio should be greater than three. Most require some skill in terms of maintenance of the probe-to-surface angle, constant contact pressure, etc. during use. There have been claims that eddy current instruments can be used for crack height measurement, but comparative trials on cracks in the range 1–15 mm high have not been promising when compared with alternative techniques such as a.c. potential drop (see Section 6.10).

Eddy current instruments need to be calibrated, and if this is done in the usual way with narrow, machined slots, it has been found that the calibration is not directly applicable to natural cracks which may be much narrower (Fig. 6.18(a)).

More promising results on the specimens used to produce the curves of Fig. 6.18(a) were obtained with a double frequency instrument (160 + 570) kHz, with which the calibration curve on slots and the measured crack heights were in good agreement (Fig. 6.18(b)).

In spite of this limitation, eddy current instruments are widely used for crack detection, particularly for cracks in difficult locations, such as screw-

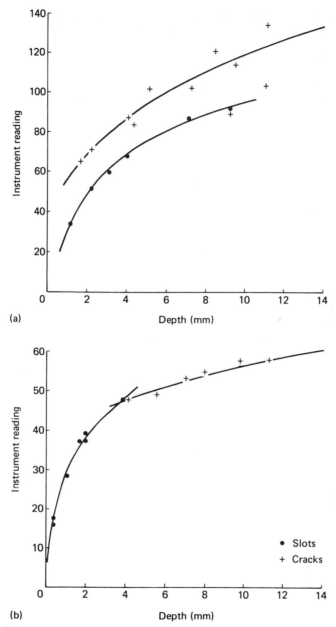

Fig. 6.18 Experimental results with two instruments for eddy current measurement of fatigue crack height compared with artificial slots: (a) austenitic steel specimens, 200 kHz and (b) austenitic steel specimens (160 + 570) kHz

thread roots, section changes or inside bore-holes. The single coil probe can be made extremely small, considerably less than 1 mm, and while the basic principle is still the change in impedance of the coil due to eddy currents generated in the specimen, modern electronic circuitry greatly improves the

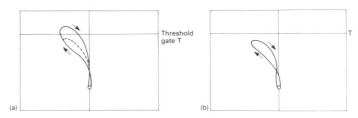

Fig. 6.19   Testing rivet-holes with rivets in place
A – cracked hole
B – crack-free hole

ability to obtain consistent and reliable results. Automatic lift-off compensation and automatic zero compensation can be built-in, together with circuitry
which automatically selects the most suitable frequency for different materials.
In aircraft inspection low frequency equipment is used to detect cracks
extending from fastener holes, without the need for fastener removal
(Fig. 6.19). Cracks in sub-surface layers as well as in the top skin can be
detected down to a depth of 10–12 mm in aluminium alloy. Using a
long-persistence screen display, a change in impedance plane pattern can be
seen as the probe is moved along a row of fasteners.

Miniaturised rotating probes can be used for bolt and fastener hole
inspection. A split-wound differential type of probe, i.e. with the receiver coil
wound in opposition, is used. High probe rotation speeds (e.g. 3000 rpm) are
used, so that a crack indication persists on the display. One limitation of this
application is that a truly laminar crack would not be detected with this type of
probe, unless the crack had an axial component. For many applications the
various eddy-current methods should be compared in performance with
flux–leakage techniques (see Chapter 5). Probe-fill factors affect flux leakage
methods more seriously, but signal analysis is straightforward.

Very-low-frequency eddy currents (10–100 MHz) have been proposed for
deep penetration applications, but the main developments appear to be in the
fields of multifrequency systems, and digital image recording and analysis.

6.9   METAL SORTING

By having two separate, identical, encircling coils, one of which is placed over
a standard specimen, any variation between the standard and a test specimen
can be detected and displayed. Some instruments have a meter display only,
but others give a more comprehensive oscilloscope display. The Foerster
'Magnatest' instruments use phase discrimination to eliminate variables such as
diameter.

In the Magnatest Q instrument, the two coil assemblies are arranged at right
angles to one another, so that the flux through one pair of coils does not pass
through the other, and both secondary coils are connected to the Y-plates of a
CRO, with a timebase on the X-plates. A 50 Hz supply is connected to each
primary coil with the two currents 180° out of phase. When the two signals are
superimposed on one another, with no test sample in the coil, the phases cancel
out and a horizontal line is shown on the CRO screen. When a ferromagnetic
specimen is placed in one of the coils, the material exhibits magnetic hysteresis

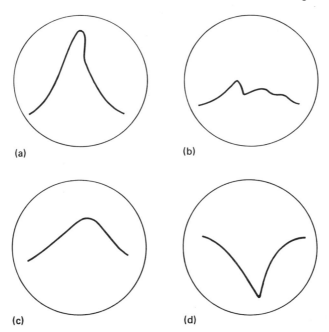

Fig. 6.20 Magnatest Q displays: (a) bar steel, hardened, (b) bar steel, unhardened, (c) chrome steel, recrystallised coarse-grained, and (d) chrome steel, recrystallised fine-grained

and the shape of the loop is modified by the induced eddy currents, so that the trace on the screen becomes characteristic of the combined effects of the electrical conductivity, relative permeability, and diameter of the specimen. If an identical specimen is placed in the second coil, the trace again becomes a horizontal straight line. The equipment can therefore be used to compare materials, or the shape of the trace can be used to identify materials (Fig. 6.20) if the specimens are identical in size and shape. Similar instruments use laid-on, instead of encircling, coils.

A large number of metallurgical and engineering properties can be measured or monitored with eddy current testing equipment. Examples are:

(1) permeability of stainless steel (to monitor cold working),
(2) cladding thickness measurement,
(3) electrical conductivity (to monitor the heat-treatment of aluminium alloys),
(4) carburisation and case-hardening depth measurement (using a multifrequency technique).

6.10 POTENTIAL-DROP METHODS

If a pair of contact points is applied to the surface of a specimen carrying a current (Fig. 6.21), there will be a potential difference, p.d., between these contacts. It is reasonable to assume that if there is a surface-breaking crack extending transversely across the line joining these contacts, the potential difference will be different. This is the principle of *potential-drop crack measurement techniques*.

In a conducting metal such as iron or steel, the p.d. between two points a few millimetres apart will be very small and measurement variations are likely to be masked by local, irregular contact potentials, but, nevertheless, the method is possible, even with a d.c. system: successful instruments have been built and used. As there is no skin effect with d.c., the technique tends to be more suitable for the measurement of deep cracks ($h > 5$ mm). Usually, three spring-loaded contact points are used, one pair straddling the crack and one pair on sound metal near the crack, as shown in Fig. 6.21.

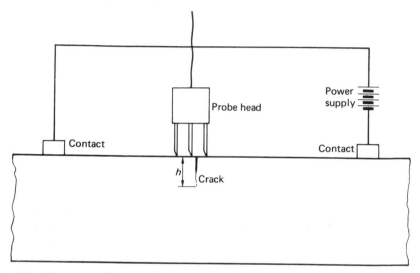

Fig. 6.21   A.c. potential-drop method of measuring crack height, h: a three-pin head is shown, but with some instruments only two pins are necessary

If an a.c. system is used, the current is carried in a thin layer at the metal surface, due to the skin effect. At a frequency of 5 kHz in mild steel with a relative permeability of 500 and an electrical conductivity of 5.8×10^6 S m^{-1}, the skin depth is only 0.13 mm. Thus, even on large structures, a current of only a few ampères is needed to give a measurable potential difference. The a.c. resistance is dependent on the relative permeability, conductivity and frequency, but if it is possible to measure the *true* surface voltage, it is possible to eliminate the need for calibration, and so eliminate many of the causes of error.

If there are two contact points a considerable distance (say 200 mm) apart on a flat surface, the current lines will spread out to give a region midway between the contact points in which the field is approximately uniform (region CDEF in Fig. 6.22). If the probe is placed on any current line in this region, but not spanning the crack, it will record the same potential difference, V_1, over the path length between the contact points, which are a distance D apart. If the probe now straddles the crack, the path length between the contact points becomes $(D+2h)$, where h is the crack height, and a second potential difference is measured, V_2. Then

$$\frac{V_1}{D} = \frac{V_2}{D+2h}$$

$$h = \left(\frac{V_2}{V_1} - 1\right)\frac{D}{2}$$

If the two potential differences, V_1 and V_2, are measured, no prior calibration of the equipment is needed. The instrument, therefore, consists essentially of an a.c. supply to set up the field, a sensitive a.c. voltmeter, and a suitable probe with two contact points a known distance apart. Stray coupling between the current supply leads and the voltmeter leads has to be avoided and because very small voltages are to be measured, the layout of the leads in relation to the specimen surface can be critical. The voltmeter needs to be of the order of 1 μV full-scale with a resolution of a few nanovolts, phase-locked to the current source. The two voltage readings can be read out digitally, or fed into a microcomputer to read the crack height directly.

On flat surfaces, where a uniform electrical field can be expected, the results of crack height measurement by the a.c. potential-drop method can be very good (for example, Fig. 6.23 shows experimental results on fatigue cracks in flat plates of austenitic steel). The instrument can be used on any conducting metal. In Fig. 6.17, there is a small, systematic deviation from the true line, for cracks deeper than 10 mm, due to the relatively small width of the cracked specimens (50 mm), which is distorting the uniform field.

This highlights one of the limitations of this type of instrument: on irregular-section specimens it may be very difficult to ensure a uniform field across the cracked area. However, this lack of uniformity can be measured and corrections made by taking a series of readings over an area adjacent to, but not across, the crack. Similarly, on short, deep cracks, a correction multiplier must be used in the equation:

$$h_2 = \frac{h_1}{M} = \left(\frac{V_2}{V_1} - 1\right)\frac{D}{2M}$$

h_1 being the measured depth and h_2 the corrected (true) depth of the crack, where M depends on the crack aspect (length/height) ratio.

As an example, for a crack of aspect ratio of 3.6, M was found to be about 0.44. This multiplier depends on probe spacing and the overall size of the crack as well as the aspect ratio, but can be calculated, and can be built in to the signal processor if the crack length is known.

Instead of making two separate voltage measurements across and near to the crack, it is possible to use a three-pronged probe, one pair of contacts being placed across the crack.

On an inclined crack, a potential-drop instrument measures the length of the crack face, P, and not the penetrated depth, h (Fig. 6.24). A forked or branched crack may give false values.

A new technique—ACFM (*a.c. field measurement*)—has been proposed, in which the surface electrical field is set up in the same way as for a.c. potential drop methods, but the voltage probe is replaced by a magnetic field probe

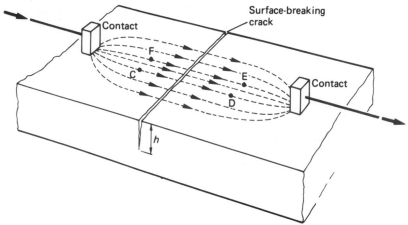

Fig. 6.22 A.c. potential-drop method of measuring crack height: field distribution on a metal plate for a crack on which the length/height is large and the contact points for the input voltage are a considerable distance apart

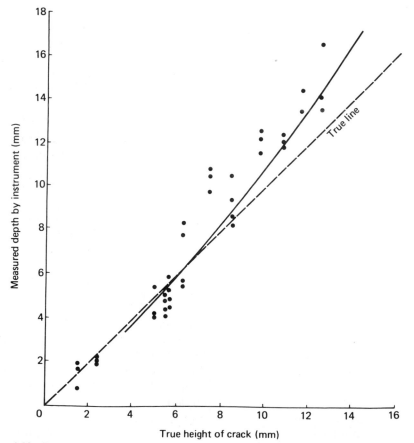

Fig. 6.23 Experimental measurements of fatigue crack height with a Microgauge a.c. potential-drop instrument (readings by different operators on the same specimens as used in Fig. 6.18)

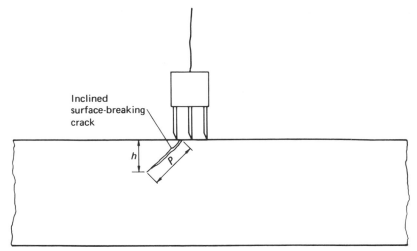

Fig. 6.24   A.c. potential-drop crack height measurement on an inclined crack

which can be electrically non-contacting. Theoretical modelling of the field can be used to size flaws without repeated experimental calibration.

The probe/surface distance needs to be accurately known but it is not necessary to have clean metal surfaces and scanning speeds can be increased. Sizing accuracies similar to those with potential-drop methods are claimed. The sensors can be multi-turn coil, magneto-restrictive or Hall-effect probes. Alternatively a probe array can be used.

The main applications of potential-drop crack measuring instruments are:

(1)   to measure the height of cracks found on the specimen surface by other inspection methods (e.g. magnetic particle flaw detection or penetrant inspection),
(2)   to monitor the height of a crack during the service life of a component.

In general, these instruments are not affected by the opening width (the separation between the crack faces) of the crack, although if there are metallic particles causing 'bridging', and these give good electrical contacts between the crack faces, false crack heights will be obtained. In typical cases of crack bridging, however, the electrical contact is not good.

The through-thickness dimension of a crack is usually of much more importance than the surface length crack because of its use in fatigue life calculations by fracture mechanics. Therefore crack height measurement on a surface-breaking crack is an extremely important application of NDT. A large number of NDT methods are used for crack detection and crack measurement, of which the a.c. potential-drop method is one of the most important.

An assessment of the usefulness of the various methods is given in Table 6.2, together with the section reference where more detail can be obtained.

6.11   OTHER ELECTRICAL METHODS

The electrical resistivity of a metal is dependent on many factors: therefore if measurements can be made of changes resulting from variations in one factor,

Table 6.2 Crack detection and crack measurement methods

Method	Section reference	Crack detection	Crack height measurement	Ferritic steel	Austenitic steel	Non-ferrous metals	Non-metals	Thin specimens (<4 mm)	Thick specimens (>100 mm)
Potential-drop									
d.c.	6.10	1	3	3	3	3	0	1	3
a.c.	6.10	1	3	3	3	3	0	2	2
Ultrasonics									
surface waves	4.2.3	2	1	2	1	1	1	2	2
transverse waves	4.6.2	2	2	2	1	1	1	1	2
0° compressional waves	4.7	1	2	2	2	1	1	0	2
TOFD	4.7	2	3	3	2	1	1	1	2
Eddy current; coil probe	6.8	2	1	1	2	2	0	2	1
Magnetic particle	5.1.3	3	0	3	0	0	0	3	3
Radiography									
film	3.8	2	1	2	2	2	2	3	1
XRTI	3.9	1	0	1	1	1	1	0	1

Code: 3 = very good
2 = moderate
1 = poor
0 = no use

resistivity measurements can be used non-destructively to monitor that parameter. The relevant parameters are crystal orientation, lattice structure, plastic strain, and imperfections, as well as porosity, cracks, and specimen dimensions.

Resistivity is of course a parameter in eddy current methods and comparatively few applications outside that field have been reported. Changes in resistivity have been used to monitor the thermal recovery of irradiated wire and for monitoring cold work at low temperatures, but these are specialised applications which appear to have no general application.

The triboelectric effect (the voltage generated by friction effects when two metal surfaces of dissimilar composition are rubbed together) has been used to detect variations in the composition of different steels, and also to sort copper-base alloys according to lead content, and lead-base alloys on antimony content.

The thermoelectric effect (the emf generated between two junctions of dissimilar metals at different temperatures) has been used as the basis of a metals comparator to detect variations in heat-treatment, metallurgical structure, and composition. A heat source is applied to one 'probe' to reach a fixed temperature, the other probe being kept at room temperature. Each probe consists of a contact on to the specimen surface made of a known alloy chosen from the thermoelectric series to give a large emf. The same instrument can be used for plating thickness measurement (copper or nickel on steel).

An electrostatic field can be used to test dielectric and insulating materials. The electrostatic field pattern is detected by an electrometer or with electrified particles. One technique of this type is 'Statiflux', used for detecting cracks in porcelain coatings. Very fine particles of chalk become electrically charged when sprayed on to the surface with an air jet, and are attracted preferentially to cracks on the surface, producing a powder build-up, analogous to magnetic particle crack detection.

Further reading: references

Eddy Current Testing
Atherton D L, Sullivan S and Daly M, *Brit J NDT*, **30(1)**, 22, 1988
Atherton D L, Czura W and Schmidt T R, *Mat Eval*, **47**, 1084, 1989
Becker R, *Brit J NDT*, **28(5)**, 286, 1986
Cecco V X S and Sharp F L, *ASM 8th Intern Conf NDE (Kissimmee)* USA, 1986
Foerster F and Breitfeld H, *Z Metall*, **45(4)**, 188, 1954
Ghent H W, *Atomic Energy Canada*, AECL-7518, 1981
Harrington R F, *Field computation by Moment Methods*, Macmillan and Co, New York, 1968
Hochschild R, *Progress in NDT*, **1**, Heywood and Co, London, 1958
Libby H L, *J Soc NDT*, **14(6)**, 12, 1956
Libby H L, *Introduction to Electronic Test Methods*, Wiley Interscience, New York, 1971
McNab A, *Brit J NDT*, **30(4)**, 249, 1988
Schmidt T R, *Mat Eval*, **42(2)**, 225, 1984
Stanford E G and Taylor H W, *Metallurgia* (Manchester), **50(79)**, 1954
Stepinski T, *Proc 4th Europ Conf NDT (London)*, 1987, **4**, 2535, Pergamon Press, Oxford, 1988

Potential Drop Methods
Dover W D and Collins R, *Brit J NDT*, **22(6)**, 291, 1980
Taylor H, Kilpatrick J M and Jolley G, *Brit J NDT*, **27(2)**, 88, 1985

7

Penetrant flaw detection

7.1 PRINCIPLES

Penetrant flaw detection methods are used for detecting surface-breaking discontinuities such as cracks, laps and folds, and can be used on any material which has a non-absorbent surface.

Many cracks in engineering materials can be deep in spite of having a very small opening width on the surface, and so can be potentially very serious defects from the point of view of their effect on the strength of the component, although they may give little surface indication. By normal visual inspection therefore, such cracks may be very difficult to detect and penetrant flaw detection is an extension of the method of visual inspection. Most of the discontinuities found by penetrant flaw detection can be seen visually if viewing conditions are perfect, but penetrant testing makes them much easier to detect.

Penetrants are solutions of coloured or fluorescent dyes in oil-based liquids. Water-washable penetrants may contain no oil. The basic principle is that a liquid which wets the surface migrates into the crack. Excess liquid on the surface is removed, and then the liquid in the crack, which carries a dye or a fluorescent substance, is drawn out by a 'developer'. The characteristics of a good penetrant liquid are therefore related to its surface tension, density and wetting properties rather than viscosity.

7.2 TECHNIQUES

The simplest and oldest penetrant test is the oil and whiting technique. A penetrating oil is applied to the surface of the specimen and allowed time to soak into any discontinuities: the excess oil is then wiped off and a thin coating of whiting (calcium carbonate powder) applied. After a period of time, the oil in a discontinuity seeps out into the whiting, causing a marked local reduction in whiteness. Hot oil is sometimes used, and also the specimen is sometimes knocked to help force the oil out.

Modern penetrant flaw detection is essentially the same process.

(1) First, the surfaces to be tested are cleaned—precleaned to remove scale, etc., degreased, and dried.

(2) The chosen penetrant is applied and a period of time allowed for it to enter any surface-breaking discontinuities.
(3) The excess penetrant is removed by a method which will not clean the penetrant out of any cracks, etc.
(4) A developer is applied.
(5) The surface is examined under appropriate viewing conditions.
(6) The surface is cleaned to prevent corrosion, etc.

The penetrant material used must be compatible with the material being inspected, and for some applications, the cleaning operation after inspection must be very thorough to prevent remaining penetrant and developer from causing contamination.

Penetrants can be classified as fluorescent, visible dye, and dual-purpose (fluorescent and visible dye).

The excess penetrant can be removed in four different ways:

(1) by water only,
(2) by a liquid solvent,
(3) by water, followed by a penetrant removal solution (detergent) which is water soluble (hydrophilic), followed by water,
(4) by an emulsifier which is oil soluble (lipophilic), followed by water.

The penetrant liquid in water-washable penetrant contains some emulsifying agent to enable the surface penetrant to be washed away, but then there is a danger of emulsifying the penetrant in the crack, so that it also is washed away. The penetrant in post-emulsifiable penetrants does not mix with water and so cannot be washed away with water alone. When the emulsifying agent is applied it diffuses only slowly into the excess surface penetrant and forms a mixture which is water-washable, but the emulsifier does not reach the penetrant in the cracks.

The developers include dry powders, aqueous liquid developers (either powder suspensions or solutions), and non-aqueous liquid developers consisting of a powder suspended in a liquid carrier.

7.3 SYSTEMS

Penetrant systems can therefore be classified as:

(1) water-washable,
(2) post-emulsifiable (i.e. the excess penetrant is emulsified for removal),
(3) solvent-removable,

and the appropriate process to be used on any specific application is based on:

(1) the flaw sensitivity required,
(2) the surface finish of the component,
(3) the compatibility of the materials with the component,
(4) the size, shape and accessibility of the region to be inspected,
(5) the ultimate use of the component.

The sensitivity of penetrant inspection processes can be very high and it is possible to detect very small cracks having opening widths of about 1 μm. In general, the more elaborate the test technique, the better is the crack sensitivity. The manufacturers of penetrant materials produce a large range of products, but there appears to be little more than qualitative comment on performance in the published literature, and often the choice depends on the compatibility of the solutions with the material to be tested, rather than on any crack sensitivity data, although reference test pieces containing cracks have been developed and are available for checking equipment performance and comparing the crack sensitivities of different processes (see Section 7.4).

Equipment available can be divided into three types, as follows.

(1)   Portable kits are used for carrying out inspection of small areas, for use on site; these often contain the materials to be used in aerosol form.
(2)   Fixed installations are used for testing components on a continuous basis, with a series of processing stations in sequential order to form a flow line. Increasingly, these are being automated to provide automatic component handling and timing.
(3)   Self-contained processing booths are used for testing large components which cannot be moved during testing.

Many of the materials used in penetrant flaw detection are a potential fire hazard and some can be toxic, so that suitable safety precautions need to be taken on any inspection installation. If ultraviolet light is used for the inspection of fluorescent indications, this also can be a hazard.

7.3.1   Precleaning

Much of the success and reliability of penetrant flaw detection depends on the thoroughness of the precleaning process. If cracks in the component are filled with debris, or access to the cracks is blocked by surface contaminants, the penetrant cannot enter the cracks and they will not be detected.

Metal-finishing processes, such as phosphating, oxidising, anodising, chromating, electrodeposition, and metallising, can obscure surface defects, and should be removed prior to penetrant testing. Penetrant methods should not be used *after* these metal-finishing procedures.

Components for penetrant flaw detection can have a wide range of surface finish, depending on the production process, and also may have been coated with corrosion preventatives, oil, grease, or paint. They may have been subject to corrosion, or have been cold worked by peening or abrasive blasting.

The first stage of precleaning is usually the removal of scale with appropriate descaling or rust-removing liquids. These generally contain inhibitors, but may be strongly acid or alkaline and residual acid or alkali can adversely affect the performance of penetrants. After descaling, therefore, the liquid must be thoroughly removed. Sometimes etching is specified after chemical cleaning.

The second stage of precleaning is degreasing followed by a detergent wash and subsequent drying in hot air.

In general, chemical methods of contaminant removal are preferred to physical methods, as the latter essentially only remove material from the surface, and may partially seal the surface discontinuities which are being sought.

Penetrant inspection is totally dependent on the ability of a penetrant to enter a surface discontinuity. If any contaminants, or the liquids used for cleaning, or any deposits produced by these liquids, fill the cracks, the flaws will not be detected. The ideal surface preparation therefore is one which leaves the surface and the flaw in a clean, dry condition. Moisture must be removed, as well as any chemical or physical residues. Moisture can be removed by heating the component above 100 °C for a period, but a preferred alternative is to immerse the component in a bath of water-displacing fluid immediately after the final washing and rinsing operations. This fluid may then be removed from the surface and from the flaw by a short exposure to solvent vapour degreasing. If the specimen is large and cannot be immersed, a volatile solvent such as ethanol or ethanediol may be used. These solvents are mutually compatible with aqueous and oily contaminants. For small components, a water-displacing chlorofluorocarbon solvent in an ultrasonic agitation tank is sometimes used, and this replaces the water-displacing/vapour-degreasing stages by a single stage.

In the aerospace industry, a high proportion of aluminium alloy components have a high-resistance paint coating, and adhesion is improved by the application of a chromating or anodised layer. If this paint has to be removed during overhaul, methods should be used which leave the surface suitable for subsequent penetrant testing. Usually, blasting methods are used, nowadays with lignocellulose powder, or with materials derived from fruit stones (plum, walnut, almond, hazel-nut, etc.). It has been found that both the precise nature of the material and the blasting pressure are critical. Loss of flaw sensitivity is probable if the blasting pressure is greater than 1.7 bar $(1.7 \times 10^5 \text{ N m}^{-2})$ on aluminium alloy components of hardness 160 VPN or less, and a light chemical etch is recommended to render the surface receptive to penetrant inspection.

7.3.2 Practical procedures

The penetrant should be applied as soon as possible after cleaning and drying, and with care that the surface temperature of the component is within the range specified for the particular material used. This may be done by brushing, immersion, flow-on, or spraying. The ideal application method provides a uniform layer of penetrant over all the surface and while immersion is the simplest process, electrostatic spray application is better for complex shapes and hollow components such as turbine blades. Whether colour contrast penetrants or fluorescent penetrants are used, there are three basic systems of water-washable, post-emulsifiable, and solvent-removable processes.

In the water-washable process, the selected penetrant is applied to the surface. The contact time varies with different proprietary products from 5 to 30 minutes, at a temperature between 10 and 35 °C. The excess penetrant is allowed to drain off, and after the appropriate contact time, the residual penetrant is removed by water rinsing. The method used will depend on the size and number of components and their surface condition, and may be:

(1) rinsing with a manually-applied water jet at low pressure,
(2) immersion in a tank with a series of fixed or rotating spray nozzles,
(3) immersion in a tank of water agitated by compressed air.

The important point is that the whole surface of every component/specimen must be covered, and there must be no spots where penetrant can be trapped. Small components are often washed in baskets. The purpose is to rinse the excess penetrant from the surface of the component without removing any penetrant retained in a defect.

After rinsing, the component is dried and it is often advantageous to immerse it in a tank of hot water for a short period to facilitate drying. The drying is preferably carried out in an air-circulating oven, thermostatically controlled to a temperature not higher than 80 °C. Drying is continued until all surface moisture is removed.

The components are then allowed to cool before applying a developer. Essentially the developer is a blotter which absorbs the entrapped penetrant which is in the flaws.

The developer can be dry powder or liquid. If liquid, it may be powder dissolved in the liquid, or a suspension and either aqueous (hydrophilic) or oil-based (lipophilic). Generally dry developers are the least sensitive, and suspensions most sensitive.

Dry powder developers consist of a white, light fluffy powder which is applied by:

(1)   applying the powder with a puffer,
(2)   applying the powder with an electrostatic powder-spray gun,
(3)   dipping the component into a container of powder,
(4)   using a fluidised bed of powder,
(5)   placing in a dust-storm cabinet.

The aim is to produce a complete, even coverage of the component, with no local pile-up of powder, and no atmospheric pollution.

The penetrant which remains in the crack provides a wet region of the surface to which the powder adheres, and then more penetrant migrates out of the crack into the developer, forming a wet mound. Because the width of the line of wet powder is self-limiting, this image spread is restricted and the crack indications tend to stay brighter and more distinct. Dry powder is usually only used with fluorescent penetrants.

With soluble developers the liquid evaporates, leaving a uniform very thin layer of white powder. Since the film is continuous, the penetrant migrates slowly and the indications gradually become indistinct. Again, soluble developers are not usually used with visible penetrants because the contrast is poor, and they are not recommended for use with water–washable fluorescent penetrants.

Suspension developers have a wider application and the concentration of powder in liquid can be adjusted to suit the penetrant used. They may be aqueous or non-aqueous. Non-aqueous developers are the most widely used and must be applied with either spray gun or in an aerosol form. It is thought that as the carrier liquid is 'wet' when applied, some of the solvent enters the flaws, dilutes the penetrant and reduces its viscosity before evaporating. Thus it is blotted out more quickly and flaw indications show up more rapidly.

The function of the developer is quite complex. With fluorescent penetrants, the light output is multiplied several times by the presence of the developer powder, by producing a microscopically rough surface on the penetrant layer, so that total internal reflection of the fluorescent light, and its consequent absorption by the walls of the crack, is prevented. The developer powder should therefore be transparent to both visible and UV light, and have a low refractive index. A similar argument applies also to red (dye) penetrants: a high-opacity powder can mask rather than enhance indications. Very fine developer powders with particle sizes less than 1 μm are not efficient enhancers of the fluorescent light.

After the developer, a period of time is allowed for the penetrant to soak out, prior to visual inspection. This time may be 5 minutes for large flaws, or up to 60 minutes for very fine flaws and maximum sensitivity. The surface inspection must take place under conditions of good illumination.

The illuminance should be not less than 500 lux (e.g. an 80 W fluorescent tube at 1 m distance), and the lighting should be arranged to avoid glare. Some means of magnification, such as a low-power lens, is advantageous.

Most colour contrast (dye) penetrants produce a red indication on a white background. Very tight cracks often produce a row of red dots rather than a continuous red line. The rate of development of the indication is often a guide to the defect size: a deep crack will produce an indication which grows and spreads with time.

With the post-emulsifier penetrants, the application procedure is the same, using the manufacturer's recommendations on contact time. The remover solution is then applied with any of the usual techniques. The contact time depends on the type of defect being sought and on the surface roughness, and needs to be determined by pre-production testing on representative specimens. On smooth, non-absorptive surfaces, a short time is used (e.g. 15 s immersion + 60 s drain time), and on rough surfaces, a longer time (e.g. 60 s immersion + 3 minutes drain time).

Excess penetrant removal can be by a number of methods. The direct use of an emulsifier (lipophilic remover) is still used in some countries, but the preferred methods are either solvent removal or the use of a solution of a hydrophilic remover. Solvent removal is widely used with dye penetrants and in automatic processes where the solvent remover can be applied as a vapour which condenses on the component and washes off the penetrant.

Hydrophilic removers (detergents) as aqueous solutions, are preceded by a pre-rinse stage to prevent rapid contamination of the remover solution, either by spray or immersion. This rinse removes most of the excess penetrant. The concentration of the hydrophilic remover solution is critical, and the maker's instructions must be followed. Different maker's products must not be mixed. The hydrophilic remover is followed by a second water rinse.

If a lipophilic emulsifier has been used, the specimens are then water rinsed to remove excess penetrant and emulsifier. A fine water spray is used to stop the emulsification and then a heavier spray used to complete the washing process. The components are then dried, developer applied, and inspected after a delay time. With lipophilic emulsifiers the time of application is critical, and if the time at this stage is extended crack sensitivity will be markedly poorer. After excess penetrant removal any of the powder developer methods may be used, or if desired a water-miscible liquid developer.

After application of the developer, a period of time between 5 minutes (for large discontinuities) and 60 minutes (for maximum sensitivity) is allowed and the specimens are then examined in UV(A) light (somtimes called 'black light'). UV(A) is the 315–400 nm region with a peak wavelength at 365 nm: the minimum desirable intensity is 50 lux (see Section 7.5) and defects appear as brilliant yellow–green indications.

The last group of penetrant processes uses a fluorescent penetrant with a solvent remover and is chiefly used on large components. The penetrant is applied with a brush or aerosol and the excess is removed by first wiping off as much as possible with clean cloths or paper tissues. Cloths or tissues are then soaked in solvent for a second wipe, after which the surface is dried and a developer is applied. For inspection, a hand-held UV lamp is usually used.

It is also possible to have a combined colour contrast, fluorescent penetrant which is first examined in daylight or artificial light and then, if higher sensitivity is required, under UV(A) light, when the flaws appear as black indications against a fluorescent blue–white background.

7.4   SENSITIVITY

Penetrant inspection is characterised by a very wide choice of materials. One manufacturer markets twenty penetrants and twelve penetrant removers, but data on the relative performance of different processes is sparse.

Fluorescent penetrant materials are capable of producing a higher flaw sensitivity than dye penetrants and some manufacturers give a 'fluorescent index' to their materials, which gives a guide to the relative performance of their own products, but there is no recognised method of rating the performance of penetrants and no means of quantitative assessment has yet been devised.

Generally, colour contrast penetrants are less vulnerable than fluorescent penetrants to contaminants, and with the latter very small quantities of contaminants such as cleaning fluids can markedly reduce sensitivity. It is claimed that the residual products from a dye penetrant examination can affect the sensitivity of a subsequent fluorescent penetrant test, and in some countries the red dye method is not used on critical components.

Water-removal processes are more suitable for routine high-volume production inspection, and for maximum sensitivity, post-emulsifier fluorescent penetrants with a hydrophilic solution remover are generally preferred. Often, the final choice of a penetrant system is made in terms of the most suitable penetrant remover related to surface finish, compatibility, flash point, etc.

Several attempts have been made to devise test specimens for comparing different penetrant processes. One proposal was a demountable block with annular mating surfaces, previously lapped together, then locally relieved by the removal of 2–10 μm. These faces were then tightened together by a control bolt. These narrow gaps, however, do not behave like natural cracks, and the lapped faces do not resemble the faces of a natural crack: the narrowest gap (2 μm) was found to be much too wide to differentiate between different penetrant systems.

A second type of test piece has been made by developing a network of cracks on a block of 2024 aluminium. By using a specific alloy, a fixed size, and a closely specified heat-treatment and quench, it is claimed that cracks of reproducible width and depth are produced. For comparison work, each block can be divided into two halves, and after use the blocks can be cleaned by vapour degreasing and stored in acetone.

Another type of test piece is a strip of steel or brass, which is chromium plated to a thickness of about 100 μm: under specified plating conditions, the plating develops a network of microcracks and it has been found that these cracks are fine enough to show clearly the differences in performance between different penetrant systems.

A set of test specimens consisting of a brass base with a nickel sub-layer and a hard chromium layer is made by the TESCO Corporation (Japan). These contain parallel cracks of depth equal to the thickness of the Ni and Cr layers (5–100 micrometres). These test-plates can be cleaned by a dilute phosphoric acid dip, followed by water spray, acetone wash and ultrasonic cleaning, then by baking at 60 °C. They can then be re-used.

A thin hard-chromium-plated sheet can be cracked by a local indentation on the back of the strip. If this is done under controlled conditions such as with an indenter on an Izod testing machine, roughly comparable star-patterns of cracks can be produced. Recent work has suggested that cracks of varying width and depth can be produced.

Some aerospace companies have compared penetrant materials and systems by producing a series of similarly cracked specimens, such as ex-service turbine blades with leading-edge cracks. The results of comparisons of different processes on these are valuable at the time of the test, but rapidly become out-of-date as newer penetrant materials are marketed.

One company has proposed a statistical analysis of penetrant testing data, using a minimum of five turbine blades having a minimum of 50 crack indications per set. The inspection procedure is standardised to match the particular penetrant process and the blades are cleaned and re-examined several times. The number of reproducible indications on two test runs with the same penetrant is then compared and a spread (distribution) from the mean value, together with a scatter coefficient, is calculated. A comparison is then made with the standard penetrant procedure. This method has the merit of checking the operator's procedure and the reproducibility of results, but does not provide real sensitivity data.

There is evidence that cracks as narrow as 0.5 micrometres in width can be detected and depth:width ratios of less than 10:1 have been claimed. The key to high sensitivity seems to be the excess penetrant removal process which is used.

7.5 VIEWING

In most penetrant inspection, the indications are viewed by eye. With colour contrast materials, very good lighting should be used (500 lux minimum) on the specimen surface. With UV light, it should be in the UV(A) band (315–400 nm), with a minimum of 50 lux.

In the past, some UV lamps have had a poor output, below this value; further, the output deteriorates with time. Instruments have been devised for measuring and checking UV illumination. Most of these measuring instruments measure the light output from a 'standard' piece of fluorescent screen, and must be calibrated, but absolute measurement instruments have also been proposed: the minimum UV(A) irradiance level at the surface of the component is recommended to be $0.5 \, \mathrm{mW \, cm^{-2}}$.

Recently, attempts have been made to automate the viewing process. The specimen is manipulated by a robotic handler, is 'seen' by a closed-circuit television (CCTV) camera, and the output signal is digitised. If this signal is then examined for any sudden changes in signal level (which would correspond to a line image) by signal differentiation, and the presence of this signal change on several successive raster lines is confirmed, an alarm signal can be activated and the specimen marked or removed for direct visual examination. Computer programs for this type of procedure are quite feasible. It is also possible to program a computer to recognise and ignore some types of spurious indication.

Early imaging systems were based on the flying-spot laser scan, but later emphasis has been on the use of closed-circuit television cameras. These will deal with quite low levels of UV illumination but many still give problems of limited depth of field. The practical problems are to ensure that all parts of the surface are 'seen', that specimen holders do not obscure or rub off indications, and that the image processing algorithms eliminate spurious indications as far as is possible. Rejected items plus a small percentage of production can be diverted to direct visual inspection.

7.6   RECORDING

Penetrant indications can obviously be photographed, or video-recorded with a CCTV camera. As with magnetic indications, with specialised methods, a fluorescent indication can be photographed to retain some identifying background. The dry indication can be lifted off the surface with transparent adhesive tape, or a replica can be made of the surface with replica-transfer-coating (RTC). This is a resin in a volatile solvent, with a white pigment and a silicone de-bonding agent. RTC is applied instead of developer, by aerosol spray, and allowed to dry. After drying, the edges are trimmed with a knife and the replica peeled off.

7.7   APPLICATIONS

Penetrant testing is used on a wide range of materials, including all metals and alloys as well as certain ceramics and plastics. Due to inadequate precleaning processes, it has had in the past a reputation for unreliability, and for ferromagnetic materials, magnetic particle inspection has usually been preferred. For small components, for production inspection, it is often automated, with a series of tanks and an inspection booth with mechanised handling, timing, solution agitation, etc. Recommendations are available for routine checking of solutions for performance and contamination. The critical factors are the cleaning process and the avoidance of contamination of all solutions. The whole inspection process is carried out with proprietary materials, and very little has been published on the physics of these materials.

Penetrants are occasionally used as a form of leak detector, by applying penetrant to one side of a specimen and a developing agent to the other side, but the overwhelming use is for crack detection on non-ferrous specimens.

Further reading

ASTM E.165, *Practice for liquid penetrant method*, ASTM, Philadelphia, USA
ASTM E.433, *Standard Reference Photographs for Penetrant Inspection*, ASTM, Philadelphia, USA
BS:6443:1984, *Method for penetrant flaw detection*, British Standards Institution, London
Lovejoy D, Penetrant Methods, Chapter 2 of *The Capabilities and Limitations of NDT*, British Inst NDT, Northampton, 1989
Non-destructive Testing Handbook, Volume 2, *Penetrants*, American Society of Non-destructive testing, Columbus, Ohio, 1985

8

Acoustic emission methods

8.1 INTRODUCTION

When a solid is subjected to stress (or environmental attack) at a sufficiently high level, sound is generated in the material and is emitted in discrete pulses. This is called 'acoustic emission, AE', or 'stress-wave emission, SWE'. The emission can be in the acoustic range, but is usually ultrasonic. The effect has been known for many years, but only relatively recently (1964) used as a method of NDT. J Kaiser (1950) is usually accepted as the originator of AE as an NDT method, and his work included many of the basic elements of modern techniques—piezoelectric sensors, high-gain electronics, etc. Some of the claims for this technique are still controversial: some authorities have claimed considerable success, others almost complete failure in detecting crack growth, etc. In retrospect, some of these failures can be traced to inadequate knowledge of the material of the test specimen, and some were due to inadequate computer capacity, so that some AE was missed.

Acoustic emissions are pulses of elastic strain energy released spontaneously during deformation, from a number of causes, such as grain boundaries sliding over one another during stressing, from plastic deformation, from inclusion cracking and from crack growth: basically, when a body suddenly deforms locally and relieves local stresses, a burst of elastic energy is emitted. Similar transient bursts of AE can be released by corrosion (chemical potential), phase changes (lattice energy) and impact (kinetic energy). External factors such as mechanical impacts, friction, machinery vibration, welding operations, can also produce acoustic emission.

These emissions then propagate through the specimen and are detected by sensors placed on the surface of the specimen, which in turn convert the energy into electrical signals. These are amplified, stored, processed and displayed. Count rates of 10^3–10^5 per minute are possible from a propagating fatigue crack, so that a large data storage capacity is necessary.

There are three basic ideas behind using acoustic emissions as a non-destructive testing method:

(1) to use several detectors and timing circuits, together with three-dimensional geometry, to locate the source of the AE, to locate where stresses are causing something to happen, such as a crack propagating. The crack is

then pin-pointed and measured by other NDT methods such as ultrasonics or potential-drop methods,

(2)   to monitor the rate of emission during stressing, to discover any sudden changes in rate which might be indicative of the formation of new defects, such as cracks,

(3)   to monitor the rate of emission and attempt to relate this to the size of a defect, such as a propagating crack, to determine the remaining safe life of a structure.

The inherent advantages of AE methods are as follows. Firstly, they are non-localised: it is not necessary to examine specific regions of a structure; a large volume can be monitored at one time. Secondly, AE can be a continuous monitoring system.

The energy release in most applications is very small, and there are usually problems in sorting out AE from extraneous noise. Acoustic emission usually falls within the frequency range 30 kHz to 30 MHz, the upper limit being set by absorption in the material, when the emission may not reach the transducer, but in most applications the high frequency emission is filtered out by a low-pass filter and the most common frequency range used is 100–300 kHz.

The emissions may be effectively continuous, or in the form of individual or short bursts of pulses (spikes, hits) of different amplitudes (Fig. 8.1). These pulses are wide-band emission, as in any background noise.

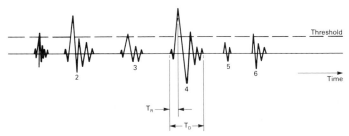

Fig. 8.1   Typical acoustic emission pulses, plotted against time.
T_R – rise time
T_D – pulse duration time
Using the threshold level, as shown, pulse 1 would register as one pulse, but pulse 4 would register as two pulses, and pulse 5 would not register

8.2   EQUIPMENT

Modern AE equipment is shown in block form in Fig. 8.2, with five channels feeding into a data store. Many more channels may be used. A typical sensor construction is shown in Fig. 8.3. Most sensors use an artificial piezoelectric material such as barium titanate ($BaTiO_3$), lead zirconate titanate (PZT) or lead metaniobate ($PbNb_2O_6$). These are sintered polycrystalline materials with ferroelectric properties which are made piezoelectric by applying a large d.c. voltage to align the dipoles. The sensor, for maximum sensitivity, may have a resonance at some frequency between about 30 kHz and 400 kHz, or it may be broad-band. By using a conical piezoelectric element and a small contact area it

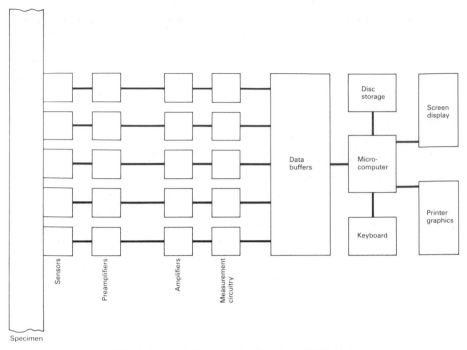

Fig. 8.2 Block diagram of a five-channel AE device

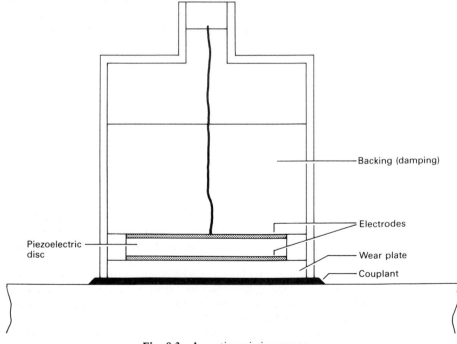

Fig. 8.3 Acoustic emission sensor

is possible to construct a broad–band sensor with a flat frequency response from 50 kHz to 1 MHz, with no resonances. A broad–band sensor is needed if frequency spectrum analysis is to be used.

Sheet sensors can be made, using polyvinylidene fluoride (PVDF or PVF$_2$) which are low cost and wide-band, with high internal damping, but low sensitivity. Sensors with some built-in electronics are also possible.

The preamplifier will have a gain of around 40 dB, with another 40–60 dB in the main amplifier, with low-pass filters on both amplifiers. Modern equipment digitises the data and has facilities for digital signal processing. Typical high specification equipment will have a 32 bit microcomputer with a 5 megabytes memory and 100 megabytes storage capacity. Parallel processing is usually necessary to handle the large mass of data. The sensor is connected to the counting and display circuitry by special low-interference co-axial cables with multiple-layer shielding, to produce immunity from electromagnetic interference. This allows long cable lengths, up to 100 m, between transducer and preamplifier. The signals, after amplification and filtration, are usually handled as digital data, but can also be displayed on a CRT.

Because of the very large range of amplitudes, amplifiers with extremely low inherent noise levels are necessary.

8.2.1 Pulse counting

The emissions, as received at the transducer, can be recorded in a number of ways

(1)   counted in relation to time: a typical method is to count the number of pulses with an amplitude above a preset value, V_1 say (Fig. 8.1)—this is called 'ring-down' counting, and is not a true count of the number of pulses, as a large pulse may be counted several times, depending on the value of V_1—thus a total count with time and a count-rate are obtained,
(2)   each counted once, i.e. discrete counting,
(3)   assessed for energy content by taking the square of the amplitude for each pulse, or measuring the area under the envelope of the amplitude–time curve,
(4)   analysed for amplitude distribution of the peak value of each emission,
(5)   analysed for frequency content (Fourier analysis or spectrum analysis),
(6)   the signal rise-time can also provide valuable information on the source of the emission.

In the train of six pulses in Fig. 8.1, if only large pulses are counted, the count would be two, but with 'ring-down' counting with the threshold as shown, the count would be six.

When there is AE with a high background count, as for example in composite testing (q.v.), the average signal duration may be nearly equal to the equipment dead-time and this can cause errors in the measurement of event duration and AE energy.

Graphically, there are two common methods of presenting AE data. Figure 8.4 (a and b) shows AE as counts/min plotted against time for an increasing load, for steel and aluminium. The other method is to plot total AE counts against time for an increasing load (Fig. 8.5).

A major practical problem in most AE work is handling the enormous amount of data produced when several transducers are producing information

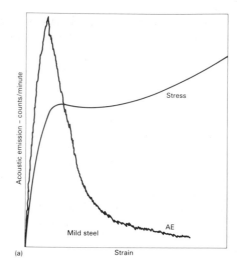

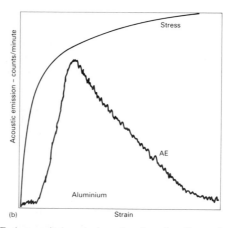

Fig. 8.4   A.E. (counts/minute) plotted against time for steel and aluminium

on a volume of specimen containing several AE sources.

It is common practice to compare or calibrate transducers by what is called the use of a Hsu–Neilson source. This consists of breaking a 0.5 mm 2H propelling-pencil-lead under a standardised procedure and recording the resulting AE. Changing the length of the pencil-lead, or varying the angle on to the specimen surface, has been found to have very little effect on the received signal amplitude.

Another proposed method of calibration of AE sensors involves the fracture of a 0.2 mm diameter glass capillary between the tip of a ball-ended screw and the flat surface of a large steel block. The sensor to be calibrated is placed at 100 mm distance on the block from the break point. The same proposal describes a 'standard' sensor against which other sensors may be calibrated.

An electric spark technique has also been used for calibration and a helium gas jet technique has been proposed.

AE simulators producing a controlled signal with known rise and decay-times, adjustable amplitude and waveform, are also available for equipment calibration.

8.3   THE KAISER EFFECT

A fundamental effect which is of major importance in AE is the Kaiser Effect. This states that if a crystal of a material is stressed while the emission is being monitored, then the stress is relaxed, then reapplied, no new emissions will occur until the previous maximum stress has been exceeded. This effect is not universal, but has been observed in both metals and composites. Some alloys and materials do not show the Kaiser effect.

Fowler has described a modified Kaiser Effect called the 'felicity effect', where emissions occur on reapplying the stress, at a specific fraction of the previous maximum load. The fraction of the load at which emission recurs is called the felicity ratio:

$$\text{felicity ratio} = \frac{\text{stress at onset of AE}}{\text{previous maximum stress}}$$

Materials where time-dependent effects control deformation, such as fibre-reinforced plastic composites, in which some of the most important applications of AE testing occur, show the felicity effect. Another example is when the AE originates from friction between free, damaged, surfaces.

8.4   SIGNAL ANALYSIS

It has been claimed that the frequency content (shape) of the AE pulse can give some information on the nature of the source of the emission, but it must be remembered that the pulse may be greatly modified in shape in travelling from the source to the detector. The form of the wave may also be changed by mode conversions, or plate thickness effects. Most AE sources generate a wide range of amplitudes and frequencies. If it is desired to study the shape of the emission pulses, as distinct from their amplitudes, it is necessary to use a broad-band transducer with a frequency range into the megahertz region. A capacitance transducer, in which the capacitance varies due to changes in the plate separation, can be used: these can have a flat response up to 10 MHz, but have a low sensitivity. The capacitor can be air- or plastic-filled.

8.5   SOURCE LOCATION TECHNIQUES

One widely-used application of AE is for flaw location. A typical application is to monitor the emission from a pressure vessel during hydrostatic testing. The method requires the use of several transducers placed on the vessel or structure and their outputs must be correlated in time to establish the position of the source of emission. There are two basic problems:

(1) to measure the difference in time of arrival of a pulse, at a pair of transducers, with sufficient accuracy,

(2)   to recognise the same pulse when it has travelled different distances, in different directions, in the material, and suffered attenuation and distortion.

Typically, 10–20 transducers, randomly located, will be used on a structure such as a pressure vessel, and experiments have shown that flaw location accuracies in the range ±3–10 cm are possible, depending on the size of the specimen. Some of the inaccuracy results from errors in the timing of the signal. The pulse is not sharp on arrival at the transducer because it contains many wave modes, and so, with timing based on the arrival of the leading edge, there is an inherent error which must be added on to instrumental errors. For a 100 cm emitter-to-transducer distance, the total error has been estimated to be about 20 μs, corresponding to a distance of just over 1 cm. If it is possible to perform the timing at the arrival of the peak pulse height instead of the leading edge, some of the error can be eliminated. Cross-correlation of the signals to determine the time-interval between two signals is also possible, but is complicated and slow and seems unlikely to lead to greater accuracy.

On complex shapes 'shadowing' effects are possible where there is no direct path for the AE from the source of the transducer. In general, it is desirable to place transducers in positions where there is unlikely to be shadowing and also where attenuation effects are minimised. True attenuation in engineering steels can be very small (<0.2 dB m^{-1}), but it is frequency dependent and so changes the pulse shape.

Signal recognition can be a difficult problem, particularly when there is copious AE and background noise. The speed of signal rise, which is very rapid for an emission event, often enables a signal to be recognised from noise, but the best criterion is when the required signal is preceded by a quiescent period. The third recognition parameter is signal length.

8.6   APPLICATIONS TO METAL STRUCTURES

It has now become widespread practice to use AE during a hydrotest of a pressure vessel. Typically the proof pressure of a vessel is 150% of the designed working pressure, and this is likely to cause some local yielding and stress relief. The vessel is loaded several times to proof pressure and, due to the Kaiser effect, if no damage has been done on the first proof there should be no AE on subsequent loadings. If there are flaws causing significant damage, there should be AE on subsequent loadings.

During a hydrostatic test AE can also be used to detect leaks. With liquids, the AE consists of background noise and small discrete pulses, and is caused by turbulence at the leakage hole. In a very large storage tank AE has been estimated to be one hundred times more sensitive than hydrostatic testing for leak detection. Visual access to the site of the leak is not necessary.

There are a number of ASME and ASTM Standards on acoustic emission methods and equipment, including ASME/E00096 'Proposed Standard for Acoustic Emission Examination during Application of Pressure' (1975). The ASME Pressure Vessel and Boiler Code, Section 8 (1985), Case 1968 cites an example where AE can be used in place of radiography and includes AE acceptance criteria for the first and second pressurisation cycles. This case is written for a small vessel (7 c.ft. 62 mm wall).

Sufficient tests have been carried out on experimental vessels, some with deliberately introduced defects, to build up confidence in this technique, but there have been a few cases where failure has occurred from a flaw which was not one of those producing large quantities of AE during the test. It appears to be possible for one AE source to increase in emission with time, and then stop, to be replaced by a second AE source which is the eventual cause of failure. Presumably, the first source extends in size, but is then relaxed and ceases to propagate: a possible example might be a slag inclusion cracking.

AE equipment is also used for 'in-service surveillance', and several permanent installations on such items as blast-furnace roofs, chemical plant, and aeroplane wingboxes have been reported.

On the application to stress corrosion cracking in hot blast stove domes, a lattice of 32 sensors was used, and many defects were detected, which on ultrasonic investigation, were found to be propagating. Location accuracies of 20 cm minimum were claimed with sensors at 5–6 m distance. A considerable factor in the success of this application was claimed to be the use of a very-low-noise front-end system, with noise levels less than 10 μV (peak).

The success of these applications is difficult to assess, as the structures are not normally allowed to operate to failure: when there is a marked increase in AE remedial action is taken. The AE records have pin-pointed serious defects, before failure. Again, there have been cases reported, particularly on structures built from so-called 'quiet' steels where the AE was low, and remained so, almost to the failure point. Some ductile, tough, low-strength steels are 'quiet' and very little AE is emitted by a propagating crack until almost at the point of failure (see Fig. 8.5(b)). It is essential to study the AE characteristics of each individual metal on laboratory test-pieces before applying AE testing to a vessel or structure of that material. Steels of similar composition can vary widely in their acoustic emission characteristics.

AE has been used with considerable success to monitor for leaks in items such as pressurised components on power plant. There have been advances in attaching transducers to surfaces with stainless steel waveguides, so as to operate in hostile environments. There is a slight coupling loss, but as this applies equally to the acoustic noise, it is not a serious drawback.

The detection of in-service stress corrosion cracking appears to be a relatively straightforward application: stress corrosion produces copious AE, and several applications have been reported.

AE has been used to detect defects in underground pipelines and some successes have been claimed. Success appears to depend on the particular material of the pipe and it seems essential to make preliminary tests on the specific material, fabricated under the same conditions (heat-treatment, stress-relieving, etc.), with either real or introduced defects. Some materials are 'quiet' almost up to the point of yielding.

Process monitoring—cutting, grinding, forming, curing—all cause AE, and so AE monitoring can be used either to characterise the process, or detect abnormalities in situ after the process completion. Examples are chatter detection in end-milling, and surface-burn in grinding, and by using a so-called 'intelligent system', a feed-back loop can be used to stop the equipment if a pre-chosen AE threshold is exceeded. There is of course a prior need to show that the AE sensitivity is related to the effect being monitored. Many metals produce much more AE than most steels and there have been many successful applications in the non-ferrous field. Deformation and cracking have been detected in zirconium alloys, due to the formation of brittle hydrides and corrosion.

8.7 DEFECT CHARACTERISATION

It is sometimes claimed that AE can provide information on the nature of a defect, and much work has been done in attempting to relate AE to metallurgical conditions. It has been claimed that the amplitude distribution (taking the amplitude of the peak in each emission) can be used as a criterion of the defect type. It is now generally accepted that in steel the predominant source of AE is plastic yield, which is enhanced at regions of stress concentration. The emission activity of steel during plastic yielding does, however, seem to be very variable: it is sensitive to sulphur content and to microstructure, and there is as yet inadequate knowledge of which steels are quiet, and which are noisy. Figure 8.5(a) shows the desirable characteristics of AE, and Fig. 8.5(b) shows what has sometimes occurred experimentally with a quiet steel.

In welds where there is a weld metal microstructure, a heat-affected zone, and a parent metal, each with a different grain structure, the AE can be markedly different on two nominally-similar welds. In one weld, a crack formed in the heat-affected zone was arrested in the refined grain zone, so that there was a dormant period in the emission immediately before failure; in a second weld, this did not happen.

It has been argued that AE can be divided into primary and secondary emissions, the latter being higher frequency and lower amplitude, requiring higher-sensitivity equipment for detection. For example, the emission from a rupture such as a brittle crack will consist of a few, very large emissions, whereas from slow, ductile crack growth the emissions will consist of low–amplitude pulses. If the emission due to the background is high, or the equipment sensitivity is inadequate, the latter may be missed completely, producing the impression of a quiet material. Cases have been reported on aluminium alloy welds where there has been visible crack extension, but no apparent increase in AE.

It is known that AE has a close relationship with material behaviour at the tip of a crack under load, so that it seems possible to obtain a correlation between AE and fracture toughness parameters (Fig. 8.6). It appears that there is a rapid increase of AE activity when the stress intensity factor, K, at the crack tip approaches the fracture toughness, K_{1c}, of the material, presumably at the point where crack growth begins. Prior to this, with tough, non–brittle materials, there is a plateau region in which the AE is broadly constant (region B in Fig. 8.6).

A number of models have been developed to describe the acoustic emission production during plastic yield with a growing crack, based on the assumption that the source of emissions is in the immediate vicinity of the elastic/plastic interface. Two different models have led to the relationships that the total emission count is proportional to K^2 or K^4. Experimental work has produced values of K^n, where $n = 1–6$. Attempts have also been made to relate AE counts to crack opening displacement values, in conditions where general

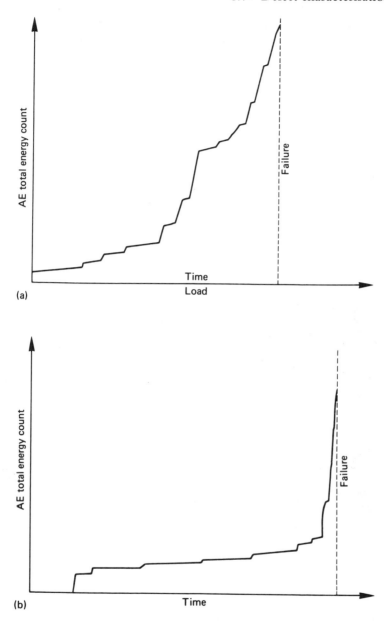

Fig. 8.5 Acoustic emission: (a) 'noisy' steel and (b) 'quiet' steel

plastic yielding is taking place, rather than brittle fracture, but there is a large variation in experimental results, and the present models do not appear to be realistic for correlating AE and stress conditions.

In relation to in-service testing of structures subject to fatigue failure, early models suggesting that the AE rates vary in proportion to the energy of new metal surfaces formed during fatigue growth do not seem to be true. There

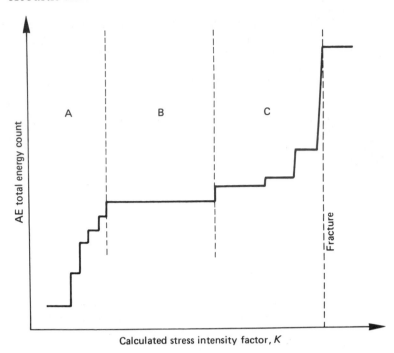

Fig. 8.6 Correlation of acoustic emission energy and fracture toughness

appears to be a logarithmic relationship between counts/cycle and stress intensity range, and this leads to the conclusion that the total number of events, during cyclic loading in which a fatigue crack is created, is proportional to the area of the crack. This has led to hypotheses on the precise mechanisms causing AE.

(1) yielding at the edge of the plastic zone,
(2) microfracture in the region very close to the crack tip,
(3) mechanical processes near the crack tip,
(4) crack growth at the maximum load point.

The mechanical processes near the crack tip will depend on the environmental conditions at the crack tip (air, water, gas, salt-water, etc.).

At present, it appears that it is possible to rely on AE methods for fatigue crack monitoring in steels only if the material involved has been studied in the laboratory under realistic operating conditions, prior to the full-scale testing. For example, three pressure vessel steels, HY80, HY100, and HY130, produced virtually no AE between yield and failure, yet steel containing MnS inclusions gave a high level of AE activity, particularly when the inclusions were oriented perpendicular to the stress axis.

8.8   APPLICATION OF AE TO NON-METALS

Glass-fibre reinforced plastic (GFRP or GRP) and fibre-reinforced plastic (FRP) are now widely used in aerospace for aircraft components, for rocket pressure vessels, and for some chemical plants. Carbon-fibre reinforced plastic (CFRP) in particular has many applications. The difficulty of producing meaningful results with conventional NDT methods on these materials has led to many investigations of the potential of AE methods.

All fibre reinforced plastics are inherently acoustically noisy. In general AE can originate from:

— fibre cracking and debonding: fibre, plastic deformation;
— matrix cracking and debonding: matrix, plastic deformation;
— interlaminar debonding (separation of two plies);
— fibre/matrix rubbing.

Failure, however, is not usually the result of the propagation of a single crack, but a progressive development of cracking, local delamination, debonding, and fibre breakage. The biggest instrumental problem is the elimination of extraneous noise, and most workers use a low-band-pass filter with the top frequency within the range 100–300 kHz.

At present, the usual method of application of AE testing to FRP rocket-motor cases is during hydrotest, and this includes a rising load, a hold period, and then a reducing load. A series of these cycles at 20%, 40%, 60%, 80%, and 100% proof load is followed and the cumulative total AE count is recorded. A typical record for a good-quality vessel for one cycle is as shown in Fig. 8.7(a) with the load curve superimposed, and two less desirable results in Fig. 8.7(b). The broken curve in Fig. 8.7(b) shows a failure before the load cycle is completed.

The duration of an acoustic event and the amplitude are both indicators of serious damage. Some workers have claimed that there are distinct amplitude regions corresponding to fibre breakage, matrix cracking, and delamination.

The further development of computer analysis of AE signals has enabled AE equipment to distinguish these differences. The practical problems are non-uniform absorption between the AE source and the sensor (most composites are anisotropic), scatter and reflections. In the absence of extraneous noise, a low-frequency broad-band sensor is ideal, but this situation is rarely met.

In airframe inspection with composite components, the main practical problem is again extraneous noise from rubbing and fretting, so that slow crack growth is unlikely to be detected against a noisy background with perhaps a hundred times more signals per unit time than are produced by the crack. One simple application was to the vertical stabiliser of an aircraft where corrosion was being caused by water ingress. By heating the area with a hot-air gun the rate of corrosion was increased to produce measurable AE, and this has become a routine test.

In many FRP materials, there appears to be a partial breakdown of the Kaiser effect, in that on a second application of load, some emission is generated before the previously applied load level is attained—the Felicity effect. The Felicity ratio has been claimed to be a measure of damage severity. With damaged material, emission has been found to continue during unloading, probably due to friction at newly-formed fracture surfaces, and this effect also serves as a damage severity criterion.

AE has been successfully used for testing adhesively-bonded joints. The amount of work reported has been limited, but it seems essential, as with steels, that pre-production laboratory experiments must be performed with

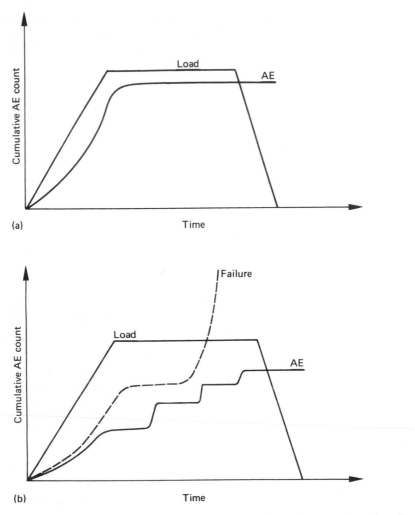

Fig. 8.7 Acoustic emission from FRP rocket-motor cases, with load curves: (a) good-quality material and (b) poor-quality material

the specific bonding materials and procedures. Some adhesives appear to be poor generators of AE.

AE has also been used for paint-testing to study the effects of environment and has been found to be a very sensitive indicator of changes in the properties of the paint-film substrate. Early signs of paint failure are easily detected by increases in AE.

Concrete is a possible application for AE testing, particularly for research into concrete compositions and for testing stressed structures such as high-alumina-cement reinforced beams. Concrete is quite different from a metallic material and is more analogous to a composite or a FRP structure. In AE measurements on test cubes, the emission was found to rise more rapidly with loading, at first linearly, and then exponentially. In the linear region the AE is low, but in the second region it rises to high values. The latter region corresponds to the onset of cracking and a departure from elastic behaviour. This exponential rise was found to commence at about 60% of failure load, with a further sudden increase in AE at about 90%. Similar effects have been found with concrete beams. AE commences, effectively at about 60% of the ultimate load, and cracking appears to commence at about 70%. As with FRP, it would seem that the Felicity effect occurs and that the Felicity ratio might be a guide to structural damage severity, but concrete is an inherently variable material and many more tests than have been reported are necessary before any firm conclusions are possible.

Wire ropes constitute another complex type of construction to which AE has been applied. Failure of individual wires can be detected in the presence of background noise, and there is an excellent correlation between wire breaks and AE events occupying given amplitude ranges. Most of the AE information was obtained by amplitude distribution analysis, the high-amplitude (80 dB) events correlating with the degradation of constituent wires. Linear source location allows break points to be located with reasonable accuracy and breaks were detected at up to 30 m distance in the particular wire ropes used (12 mm diameter 7 × 7-strand steel core construction).

No published work has been traced which compares the relative performances of AE, electromagnetic, and magnetic methods for the NDT of wire ropes.

AE has found applications outside the conventional NDT field in biomedical engineering, in glacier studies, in earthquake studies, but one interesting new field is PIND (particle impact noise detection) in electronic components. This uses AE to detect loose particles during vibration or shock loading. Particles as small as 0.16 microgram can be detected, and there is a US MIL-Standard 883C; Method 2020 on the subject. There have been several successful applications of AE in process plants, such as the mixing of dry powders, where the AE becomes constant when the powders are fully mixed.

8.9   ADVANTAGES AND DISADVANTAGES OF AE METHODS

AE is one of the newest NDT methods but has now passed through most stages of the 'life-cycle' of a new method:

(1)   discovery—exaggerated claims;
(2)   failures due to overselling of the method;
(3)   basic studies; accumulation of data;
(4)   development of reliable techniques;
(5)   general acceptance and production of standards.

There are about 15 American Standards on AE, several of which are specific to particular applications and which contain acceptability criteria. Perhaps the most important standard amongst these is ASTM.E.750 'Measuring the Operating Characteristics of AE Instrumentation'. The European Working Group on AE has issued a number of Codes of Practice. There are three Japanese Standards on AE.

The advantages of AE are:

(1)   It gives large area coverage.
(2)   It details growing discontinuities.
(3)   It locates discontinuities.
(4)   It can detect discontinuities in inaccessible areas.
(5)   It can be used as an on-line test.

The disadvantages are:

(1)   Not all discontinuities emit detectable AE.
(2)   It depends on the method of loading the specimen.
(3)   Signals can be obscured from the sensor by geometry, noise or absorption.
(4)   It does not determine the size of the discontinuity.
(5)   There is no standardised data interpretation.

8.10   FUTURE DEVELOPMENTS

The use of computer programs to analyse the AE will increase rapidly with increases in computer capacity. Data analysis to determine correlated parameters and assessment of the best discriminating parameters has already progressed well beyond classical statistical techniques. Programs such as 'Principal Component Analysis' are already in limited use (Roget, 1988). Modelling of AE will improve our understanding of the form of the emissions. New designs of sensor with acoustic waveguides, rolling contact, integral electronics are already in development.

Further reading: references

Bar-Cohen Y, A review of NDE of fibre-reinforced composite materials, *Mat Eval*, **44(4)**, 446
De Meester P J A and Wevers M G T, NDT of composites, *Proc 4th Europ Conf NDT (London)*, **1**, 95, Pergamon Press, Oxford, 1988
Fowler J T, *Progress in Acoustic Emission*, Yamaguchi Y (ed), **3**, 150, Jap Soc NDT, Tokyo, 1986
Kaiser J, *PhD Thesis*, Tech Hochschule, Munich, 1950
Kaiser J, *Arch Eisenhuttenwesen*, **24**, 43, 1953
Non-destructive Testing Handbook (2E), Volume 5, *Acoustic emission testing*, Miller R K and McIntire P (eds), American Society of NDT, Columbus, Ohio, 1987
Proc 4th Europ Conf NDT (London), **4**, 2849–3100, Pergamon Press, 1988
Roget J, *Proc 4th Europ Conf NDT (London)*, **4**, 3083, Pergamon Press, 1988
Segal E and Rose J L, NDT of adhesive bonded joints, in *Research Techniques in NDT*, Volume 4, Sharpe R S (ed), Academic Press, 1980
Summerscales J (ed), *NDT of Fibre-reinforced Plastics Composites*, Elsevier Applied Science Publishers, London, 1987
Williams R V, *Acoustic Emission*, A Hilger, Bristol, 1980

9

Other methods

Almost any physical measurement can be converted into a non-destructive test applicable to some particular problem or material, and it is extremely difficult to predict when a very specialised esoteric technique might turn into an important new NDT method.

If a measurement of a particular physical property is affected by flaws in the specimen or by a change in metallurgical structure which affects the bulk strength of the specimen, measurement of that property is a form of NDT. The main difficulty is in establishing a relationship between a particular measurement and a particular mechanical property without another property of the material completely swamping the results. Thus in some metals an apparent relationship might be established between ultrasonic attenuation and tensile strength, but this relationship is masked by the effects of varying grain size and other metallurgical parameters.

A distinction can sometimes be made in terms of whether the measurements can be made on a bulk specimen. If it is necessary to cut or prepare a sample for measurement, this generally is regarded as outside the field of non-destructive testing.

Such methods as positron annihilation, microwaves, magneto-reflection, exo-electron emission, radiowave polarisation appear at the moment to be of limited interest, although all these methods and many others have been reported as being used for some particular application of NDT. Such techniques as thermography, sonic methods and holography have, however, already become well-established.

This chapter therefore will cover only those methods which already have a fairly wide usage, and will not attempt to cover the more esoteric techniques. In Eastern Europe use seems to be made of rather obscure magnetic and electromagnetic properties to monitor the properties of materials and many theoretically-based papers have been published. These techniques do not seem to have caused much interest elsewhere and little has been published on comparisons with better-known NDT methods. Some of these lesser-known methods, such as the use of Barkhausen noise, have been briefly outlined in Chapters 5 and 6.

9.1   ACOUSTIC METHODS

Acoustic methods of NDT go back to the railway wheel tapper, his hammer and his ears. There have been frequent claims that the process could be automated—a standard 'tap'—and the signal analysed electronically, to provide useful information on the quality of the specimen. Unfortunately, as in much NDT, there are rather a lot of variables—specimen shape and size, metallurgical structure, presence of macro-defects—and some doubts as to the best properties to measure in the ensuing pulse, e.g. decay-time, frequency content, ringing-time, amplitude, etc. Even so, in recent years the coin test for composites has proved to be a good practical solution to the problem of detecting local defective areas. Each point on a structure is tapped and defective areas have been found to produce a 'dead' response. It is not really analogous to the wheel-tapping since it is a local area, not a global test. The sound produced when a structure is tapped is mainly at the frequencies of the mean structural modes of vibration, so the sounds produced when a good area and a defective area are tapped must therefore be due to a change in nature of the force input. By performing a Fourier Transform on the force-time records, the frequency content of the signal can be obtained. Pulses can be compared in both the time or frequency domains. Delaminations in solid CFRP and crushed cores in honeycomb structures can be found. The sensitivity decreases with increasing flaw depth and a flaw 10 mm diam. under a 1 mm thick CFRP skin is near the limit of sensitivity (Cawley, 1989). The chief advantage of the method is that no couplant is needed under the tapping head and the data required can be obtained through a transducer built into the head.

There is a range of related methods, variously known as sonic, acoustic, vibrational or mechanical methods, all depending on some measurement of mechanical vibrations.

On one particular material, sonic methods have had considerable success. This is nodular iron (sometimes called spheroidal-graphite cast iron). In cast iron, the free carbon can take a number of forms—graphite flakes in grey cast iron, or spherical nodules in malleable cast iron—and the mechanical properties of malleable iron depend partly on the matrix structure and very markedly on the graphite morphology. If the graphite has a microscopic spheroidal shape, the interface with the matrix is small, and an improved toughness and better mechanical properties are obtained. The amount of graphite spheroidisation can vary from 0% to 99%, and is one of the properties which is most difficult to control in making castings in malleable iron, as it is controlled by additions to the ladle immediately before pouring, or in the individual moulds. Malleable iron castings are widely used in the automobile industry.

An experimental correlation has been found between Young's modulus, E, and the degree of graphite spheroidisation: therefore, as the velocity of elastic waves under broadbeam conditions, V_L, is given by

$$V_L = \left(\frac{E}{\rho}(1-v)(1+v)^{-1}(1-2v)^{-1} \right)^{1/2}, \text{ assuming } \lambda \ll D$$

where v is Poisson's ratio and ρ is the density; the degree of graphite spheroidisation should also be related to the wave velocity.

A second parameter found to be related to the spheroidisation is the elastic wave attenuation. There appears therefore to be good reason to expect that a measure of some elastic wave properties can be used to measure spheroidal-graphite content. If the transit time of an ultrasonic beam between two faces of a casting a known distance apart is measured, this is a measure of wave velocity, and if the pulse heights are measured, also of ultrasonic attenuation. These measurements, however, are difficult to make and will depend on the material properties only within the line of the ultrasonic beam.

The resonant frequency of the casting depends also on Young's modulus and will produce a value related to the average degree of graphitisation over the whole casting. The casting is suspended at one or two points, to keep damping to a minimum, and with a small attached oscillator, the resonance frequencies are found by the peak of a small receiver. The resonance frequencies can be related directly to the degree of spheroidisation, or a frequency spectrum can be produced and compared with that from a standard satisfactory casting. For a complex-shaped casting, the spectra may be quite complex.

An alternative technique is pulse excitation, with a microphone pick-up and a spectrum analyser. This method minimises the effect of environmental noise but involves difficulties of acoustic coupling. With small castings, there may be complications from the spectrum of the striker, and an alternative shock excitation technique has been developed by Magistrali.

The excitation energy is supplied by allowing the casting to drop from a given height on to a large striking block of highly-damped material. As the casting 'bounces off' the block, an omnidirectional microphone picks up the acoustic signal which is then preamplified and analysed. An alternative excitation technique is to push the casting off the end of a support table with a striking bar, and again detect the signal while the casting is in the air (Fig. 9.1). The type of signal obtained is shown in Fig. 9.2(a), and its frequency spectrum in Fig. 9.2(b). A spectrum from a grey iron casting is shown in Fig. 9.2(c).

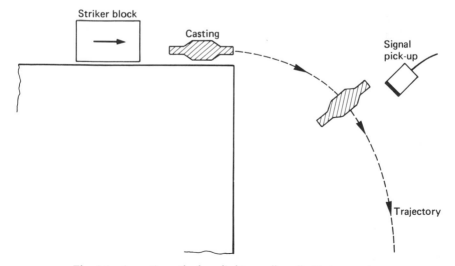

Fig. 9.1   Acoustic method applied to small, malleable iron castings

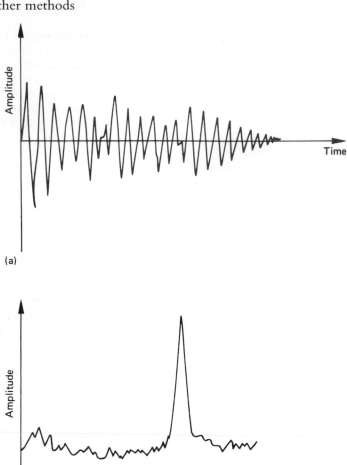

(a)

(b)

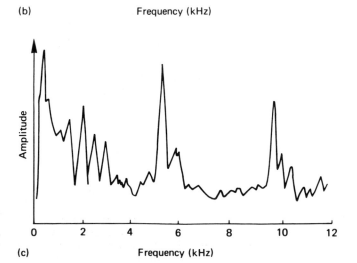

(c)

The signal used is triggered from the leading edge of the signal and only a limited length is taken, so that any noise associated with the impact is eliminated. It is then digitised, Fourier analysed in a computer and displayed. For routine inspection, an active high-pass filter can be applied with a cutoff frequency selected to allow a high gain for the resonance frequencies obtained with ductile iron castings of satisfactory quality. By this means, castings can be segregated into classes. The range of resonance frequency shift due to extraneous causes has been found to be 10–20%, for otherwise satisfactory quality castings, and the acoustic response has been found to be independent of the revolutions of the casting along its trajectory.

A technique based on the same principles has been proposed for the inspection of the bonded layer in shock absorbers. The propagation stress wave is produced by a spark discharge on to the metal surface, from an electrode: multidirectional waves of different modes are generated and detected with a wideband sensor (20 kHz to 1 MHz) and a portion of the received wave is frequency analysed. The sensor is scanned over the bonded area and a C-scan display is produced of regions where the amplitude of a particular chosen frequency is higher than a threshold value.

The presence of damage in FRP specimens can be detected by a change in stiffness, which is most easily found by a change in the natural vibration frequency. The method can easily be automated to become a very rapid sorting process for production quality control (Fig. 9.3). Acceptance limits for filament-wound GFRP tubes were set at three standard deviations from the mean value for good specimens. The sensitivity of the test was therefore

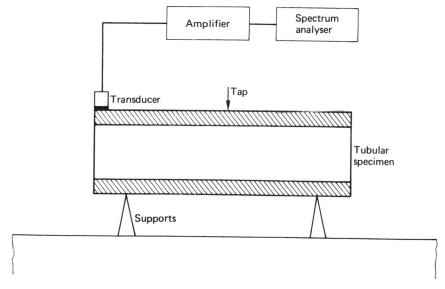

Fig. 9.3 Acoustic testing of GFRP tubular specimen

Fig. 9.2 Signals from technique shown in Fig. 9.1: (a) unprocessed signal, (b) frequency spectrum for malleable (ductile) iron casting signal, and (c) frequency spectrum from grey cast iron signal

determined by the scatter in the results from the good specimens: this variation was found to be too high to detect small, localised defects.

If the defects are to be located and assessed, however, a dynamic analysis is necessary and it may be difficult or impossible to carry out such an analysis on complex-shaped specimens. Using much lower frequencies (about 2 kHz), masonry structures have been tested by similar methods. The structure is excited by a blow with a steel-tipped hammer, and the frequency of the propagated signal analysed through a transient recorder in conjunction with a microcomputer. When the method was applied to masonry bridge structures, the signal was found to be sensitive to significant defects such as large voids.

Attempts have been made to use measurements of acoustic intensity made at points over a very large structure to monitor fatigue cracking developing in the welds of the structure.

9.2   MECHANICAL IMPEDANCE METHODS

There is a range of methods which involve generating a pulse of mechanical energy by mechanical or electrical methods over a *local* area and looking for local changes in resonance, spectral, or decay characteristics. The major application is to adhesive bond testing. Most of these methods use mechanical waves in the 1–10 kHz frequency range (sonic or ultrasonic).

The strength of an adhesive bond is governed by the two fundamental properties of adhesion and cohesion. *Adhesion* is a measure of the degree to which the adhesive is attached to the bonded surfaces, but there seems to be no basic theory of adhesion which would point to a precise physical property which could be measured. *Cohesion* is a measure of the strength of the thin layer of adhesive material and is principally affected by the adhesive thickness, its modulus and its density.

In a good bond, the adhesive strength is always higher than the cohesive strength, and when a bonded joint is pulled apart, failure should be by splitting of the adhesive layer, the adhesive remaining attached to the bonding surfaces: no bare bonded surface should be visible. In practice, in making an adhesive-bonded joint, there can be failures due to areas where there is no adhesive, lack of adhesion due to contamination of the surfaces, or poor cohesive strength due to faulty curing or porosity in the adhesive, etc.

The acoustic impedance, Z, of a material is given by

$$Z = [E_d(1 + i\eta)]\rho$$

where E_d is the dynamic Young's modulus, η is the loss factor, and ρ is the density.

The generalised elastic modulus and the acoustic impedance are complex numbers. The reflectivity and transmissivity of a mechanical wave at any interface between two media having different impedances are given by the same equations as in Chapter 4 (equations 4.5 and 4.6).

In general, the mechanical waves used in bond testing are generated by a contact probe using piezoelectric excitation, and the same probe contains the receiver crystal (Fig. 9.4). It is possible to use either a single frequency, a multiple frequency, or a swept frequency.

Most bond testers look for local standing wave resonances on a point-to-

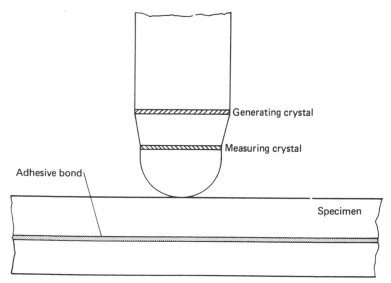

Fig. 9.4   Adhesive bond testing probe

point test, with a varying frequency, but it is also possible to obtain information from the amplitude of the received signal and its phase difference from the transmitted signal. The received signal can be digitised for filtering, spectrum analysis, etc. The reflected wave is very sensitive in both amplitude and phase to a change in impedance of the second material (the adhesive) at the interface. A useful technique with swept frequency equipment is to display the difference between spectra obtained over a good bond and the area under test.

Practically every well-known method of NDT has been tried out for the testing of adhesive bonding in laminates and honeycomb structures for aerospace applications, and overall confidence-level tables have been proposed. These must inevitably depend to some extent on operator skill and on the particular bonding defects being considered, but confidence-level values of 72% for immersion ultrasonics (C-scan, through-transmission), 53% for the Fokker Bond Tester (resonance continuous wave), to 30% for the simple tapping technique, have been reported (Segal and Rose, 1980). These values do not take account of adhesive failures.

It seems probable that the future of adhesive bond testing will lie with ultrasonic techniques using a computer for data analysis by pattern-recognition algorithms for bond strength classification, although the lower-frequency acoustic methods have the advantage that dry contact probes can be used and the positioning of the probe is not so critical, because of the longer wavelength. The method is most sensitive when the defect is close to the surface and the base structure is relatively stiff.

A more rigorous approach to some adhesive testing has been to measure the complex input mechanical impedance (i.e. the absolute value and phase of the input impedance) of a vibrating specimen, the specimen being in rod–like form.

Methods based on the damping capacity of free oscillations are also called 'internal friction' measurements. These use small amplitude vibrations, which are relatively easy to make, either as the energy input to maintain a constant amplitude vibration, or by measurement of vibration decay (as described above, for ductile iron castings), or by measurement of attenuation. It ought to be possible to monitor the effects of impurities in materials by continuous damping capacity measurements and a few laboratory investigations have been reported, on widely differing materials—e.g. carbon and hydrogen in steels, contamination in semiconductors, micro-quenching cracks in steel, irradiation effects in polythene, and carbon in nickel.

9.3   VIBRATION MONITORING

Even the most simple machinery has a number of sources of vibration, and these vibrations are transmitted through the machine structure, so that at any point on the machine there is a vibration spectrum which is the complex sum of all the individual components of the vibrations and their harmonics. This vibration spectrum will depend on machine speed, loading, lubrication, etc.

Periodic measurements can be made of a machine's vibrations and compared with earlier measurements made under identical operating conditions, to detect any significant changes which may be indicative of a deterioration in the machinery. The measurements can be continuous, online, or be periodic checks.

A vibration is characterised by amplitude, frequency and phase, and is usually measured by a small electromechanical transducer which produces an electrical output. The transducer can sense displacement, velocity or acceleration, and the accelerator type is the most widely used, being small, immune to magnetic fields, and capable of being used at high temperatures. The plane of measurement of the transducer must always be the same, so that hand-held transducers must be applied with great care.

The circuitry attached to the transducer should contain a series of filters to isolate and measure narrow bands of frequency throughout the vibration spectrum and there should be equipment for storing a full vibration signature.

With online vibration monitoring, computer controls can be programmed to shut-down, or reduce speeds, if preset limits are exceeded. Obviously, the main applications are to rotating machinery, and as machinery failures are seldom sudden, catastrophic events, online monitoring can give a long warning period of impending failure, with a possibility of a diagnosis of the causes.

Vibration monitoring has been successfully applied to such complex items as nuclear power plants, including the reactor pressure vessel. Loose parts, material under-going fatigue damage, burn-up of fuel elements, have been identified by vibration analysis.

9.4   LEAK DETECTION

Many structures are designed to be pressurised, or to be pressure-tight, and as a consequence, leak detection is an important NDT method.

Defects are revealed by the flow of gas or liquid into or out of the test specimen. The simplest and most widely-used method is the hydrostatic test, using either clean water or water containing a dye. A simple practical method is to use a fluorescent penetrant in the liquid and search with a UV lamp. The presence of leaks can be revealed by water or gas seepage, or by changes in pressure. Bubble formation when the specimen is covered with a soap solution or immersed in a liquid can also be used for defect location, an obvious example being the water-bath testing of the inner tubes of automobile tyres. The bubble test is not particularly sensitive unless very high pressures are used, which is highly dangerous because of the expansive force of gas under pressure. In water, comparatively large bubbles are formed and take a long time to appear, and a liquid with a lower surface tension can be much more sensitive. In general, search gases can provide much greater sensitivities with appropriate gas detectors, but sensitivity has to be balanced against convenience and hazards.

In the most sensitive techniques, hydrogen or helium gas is used with a mass spectrometer detector and sensitivities up to 1×10^{-7} bar s^{-1} are possible.

$$10 \text{ bars pressure} = 1 \text{ pascal} = 1 \text{ newton per metre}^2$$

The search gas used should not be present in air in any significant concentration, should preferably be inert, and should have a molecular mass not requiring critical tuning of the mass spectrometer.

Helium is now the preferred testing gas for high-sensitivity applications because of its inert nature. Typical large-scale leak testing applications are to domestic refrigerators and components, power plant steam condensers, and underground pressurised telephone cables. In the last application, because the gas has a high diffusivity, it rises to the surface in a short time through sand, earth, or concrete, and the location of the leak can be found to an accuracy of ± 15 cm at a depth of 70 cm. The tests commonly take place over a length of 20 km of buried cable at a time.

There are many methods of detecting leaks, depending on the sensitivity required and Table 9.1 gives some indication of these and their performance.

The ultrasonic detectors use acoustic-emission probes (piezoelectric elements), which are omnidirectional, require access to the outside surface only, and are comparatively cheap. The turbulent outflow at a leak produces acoustic emission which is transmitted through the material. The disadvantage of this system is that the acoustic emission from the leak must be separated from other sources of AE.

Table 9.1 Leak detection methods

Method	Probe	Minimum detectable leak (cm^3/s)	Comment
Water immersion	air/He	10^{-3}	low cost
Dye penetrant	—	10^{-4}	low cost
Pressure changes	—	10^{-5}	low cost
Ionisation gauge	H, He, O	10^{-7}	—
Ultrasonic	air	10^{-2}	—
Radioisotope K-85	counter	10^{-10}	special safety precautions needed
Mass spectrometer	He	10^{-10}	versatile; very costly
Sniffer	He	10^{-6}	versatile; costly

In the radioactive gas methods, either a direct leak of radioactive gas is detected with radiation detectors, or radioactive gas is applied, purged, and then areas where gas has lodged can be detected. The method is extremely sensitive; precautions are necessary both to prevent permanent contamination and to protect personnel from excessive radiation exposure.

Halogen gases have a strong affinity for electrons and extremely sensitive halogen gas detectors can be built for use with any gases containing chlorine, fluorine, or bromine, such as the Freon gases (SF_6).

In the heated-anode halogen detector, which is a widely-used system, the gas from the collector flows over a heated alkali emitter (the anode) and the negative halogen ions greatly increase the emission of positive alkali ions and so increase the current across the platinum electrodes.

In the electron capture gauge, a non-electron-capturing gas such as nitrogen or argon is used as a background gas, to carry any traces of halogen gas. This gas flows across the sensor element, where it is ionised to produce free electrons by a weakly radioactive source such as tritium ($_1H^3$), which decays by electron emission to $_2H^3$. The half-life of tritium is 12 years and the electron energy is 18 keV. Thus if there is a trace of halogen gas in the gas passing over the sensor, the electron is captured and the electron current is reduced. Electron capture gauges have a very good calibration stability, are not sensitive to contaminants, and can have a sensitivity as high as 10^{-12} Pa m^3 s^{-1}, using SF_6.

A similar very high sensitivity can be achieved by using the mass spectrometer as a helium leak detector. Gas molecules entering the mass spectrometer are bombarded with electrons from a heated filament (Fig. 9.5). The ion beam produced is accelerated in an electric field as a narrow beam and then deflected by a permanent magnet into circular paths of different radii, depending on the mass of the ions. A detector is arranged to be in line with the helium ions, so that only these are collected. The detector can be connected through amplifiers to a meter or an audio output. The helium leak detector can be used as a sniffer, or as a continuously sampling probe. If a specimen cannot be filled with helium, it can be placed in a pressurisation chamber containing helium for a period, removed, and then transferred to a (second) vacuum chamber to which the helium detector is attached. Any helium which has entered the specimen in the first chamber through the leakage points then flows out and is detected.

An important application of leak testing is to detect leaks in storage tanks. Vapour-phase leak detectors are used, i.e. detecting gas in air, using various detector gases such as isopentane, benzene, toluene.

9.5   THERMOGRAPHIC METHODS

Thermography is based on the measurement of the heat distribution across a surface. It may be a passive technique, i.e. the natural heat distribution is measured over the surface of a hot structure. Or an active technique may be used, where a heat pulse or a series of pulses is applied to part of the surface, and the movement and redistribution of these across the specimen is measured. With an active technique transient or stable conditions may be used. If heat pulses are generated with a modulated laser beam, periodic temperature

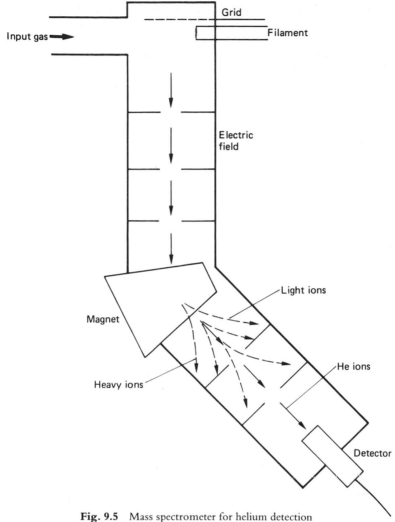

Fig. 9.5 Mass spectrometer for helium detection

variations—thermal waves—are produced in the specimen: these can undergo reflection, partial reflection or interference, and have been used to study coatings. The detection system can be a contact (surface) system such as a cholesterol liquid crystal layer, or a tele-system such as a thermographic camera. Surface systems tend to be low-cost and simple, and can have high resolution. Camera systems are more expensive.

The first industrial applications of thermographic (infrared) techniques were the measurement of stationary temperature fields, such as the measurement of temperature across hot-rolled steel strip, or variations in insulation on a wall of a building. There have been a few studies with contact sensors, but most applications use a non-contact thermographic 'camera'. With this, the infrared image of the specimen is imaged by a suitable lens on to an infrared-sensitive detector. Various mechanical and electronic scanning arrangements have been developed to build up a line-scan, and then a series of line-scans, to give a two-dimensional image on a television raster. A modern infrared camera might typically have a cadmium–mercury–telluride detector, with a spectral bandwidth of 8–14 µm, a germanium lens system, a spatial resolution of about 1 milliradian, a temperature resolution of about 0.2 K, and an effective temporal resolution of about 20 ms. Such a camera can therefore operate at normal video-rates and can be used for transient thermography. Earlier infrared cameras have used radiometer detectors with mechanical scanning, pyroelectric elements, and cooled elements (77 K), but the advent of multiele-ment arrays of detectors, which will operate at room temperature, would appear to make other systems obsolete.

The basic idea behind thermographic non-destructive testing is that if a pulse of heat is applied to one side of a specimen, the spatial distribution of the heat flux on the opposite surface will depend on the homogeneity of the specimen (i.e. the presence of internal defects), the diffusivity (which is related to the thermal conductivity and the volume heat capacity), and time. The difference between thermographic methods and other methods using electro-magnetic radiation is that the heat is propagated by a diffusion process and not instantaneously. The difference in temperature on the surface facing the detector can be considered as an 'image contrast' and can be related to the size of an internal flaw, the heat to be deposited per unit area, and the time that it takes the contrast to develop. For a sheet specimen of thickness L, the time $t_{1/2}$ for the back surface to rise to half of its ultimate temperature increase has been calculated to be

$$t_{1/2} = 1.38L^2/\pi^2\kappa$$

where κ is the thermal diffusivity, which has values of $10–100 \times 10^{-6}$ m^2 s^{-1} for metals and $0.1–4 \times 10^{-6}$ m^2 s^{-1} for GRP. Thus for 1 cm copper, $t_{1/2}$ = 0.13 s, and for GRP perpendicular to the fibres, $t_{1/2}$ = 110 s. For copper, therefore, the heat diffusion is so rapid that the contrast would be difficult to capture, but for the non-metallic material there would be no problem.

The heat pulses can be generated by a flash tube which delivers energy in very short pulses and can generate high initial temperatures at the heated surface. The practical problems are controlling the uniformity of the heating cycle, the variability of surface emissivity, and convective cooling effects producing spurious images. In spite of these problems, there have been successful applications such as the detection of delaminations and resin-rich regions in CFRP laminates, flaws in honeycomb-bonded composites (CFRP-Al), in metal-fibre adhesive-bonded composites and one-side thickness measurement in CFRP. This type of video-thermography requires no physical contact between the detector/recording camera and the specimen, and the full facilities of video-cassette recorders, image processing, etc., can be used.

Flaws smaller in diameter than about twice their depth below the surface produce only a very small thermal contrast at the surface because of the heat diffusion process. With anisotropic material such as CFRP, because the conductivity in the laminate plane is much higher than through the thickness, the contrast due to flaws is further decreased. Thus pulsed thermography is unlikely to be successful on thick specimens, or in the detection of deeply buried flaws. The transient flow from a pulsed heat source can be modelled on a finite-difference computer program, to predict the surface temperature distributions and the attainable heat–contrast values.

It would appear that further development of video-thermography is linked to the development of improvements in pulsed heat sources, with laser heaters as possibly the most important field. With modern flash tubes, energies up to $5\,\mathrm{J\,cm^{-2}}$ can be delivered in a few milliseconds, but in some materials, such as the resins in laminates, such high energies could produce physical damage and degradation, so that care must be taken on the selection of heat sources for specific tasks. Temperature rises of over $100\,°C$ can be calculated for GRP and CFRP materials, assuming an input of $1\,\mathrm{J\,cm^{-2}}$.

A number of variants in technique are possible. The heating element (spot, line, or area) can be on the same side of the specimen as the detector (Fig. 9.6), the specimen being scanned. The temperature distribution will have a bell-shaped form along the line of specimen travel and the optimum delay-time between heating and measurement needs to be calculated.

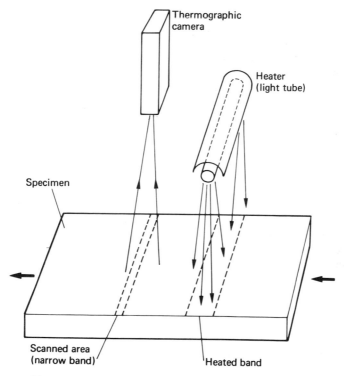

Fig. 9.6   Thermography of moving plate-shaped specimen

There will always be sensitivity limitations due to noise problems and emissivity variations. Coating the specimen surface with a uniform black layer can reduce the variation in signal to about 4%, but this is not always an acceptable technique.

9.5.1   Non–camera detectors

The surface temperature distribution on a specimen can be revealed by a cholesteric liquid crystal coating applied to the surface. Liquid crystals are grease–like chemical compounds which appear to change colour with temperature: they are relatively cheap. Such cholesteric crystals can cover a full colour range from red through yellow, green to violet in a temperature range of 4 °C, from 33–37 °C. The surface of the specimen is coated with black paint and the liquid crystals applied by spray. There is a slow degradation of the liquid crystals in air, so that testing must take place shortly after application. The resulting heat images can be recorded by colour photography, at intervals after heating.

An interesting application of thermography to small specimens is shown in Fig. 9.7. The heating is supplied by the electron beam in an SEM microscope, and the emitted infrared radiation is reflected by an ellipsoidal detector with the specimen at one focal point and a pyroelectric detector at the other. The detector is covered by a silicon window which blocks visible radiation but transmits infrared in the 10 μm wavelength region. The detector samples the entire specimen surface, as the electron beam is scanned over the specimen, measurements being made just before the beam moves step–wise to the next point. The imaging system uses a phase–sensitive narrowband receiver, and artifacts due to surface emissivity variations are cancelled out in the phase image. A faster cadmium–mercury–telluride photoconductive detector can reduce the imaging time by 10^3.

A method known as 'vibrothermography' has been applied to thin, composite structures. Vibrothermography is based on a conversion of elastic energy to thermal energy, by a rubbing action, at the location of a defect, during cyclic loading. The specimen must have a relatively low thermal conductivity

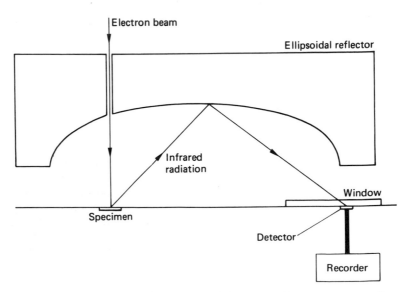

Fig. 9.7   Thermographic system using an electron beam in an electron microscope for specimen heating

and the test is best carried out under dynamic conditions, such as with real-time video-thermography.

9.5.2 Applications

The main interest in thermographic testing methods is for the testing of non-metallic materials, such as composites and laminates, where the conventional NDT techniques of radiography and ultrasonics produce results which are very difficult to interpret.

There have been many attempts to apply thermography to adhesive bonding, which is widely used in the aerospace industry. The major disadvantages have already been outlined:

(1)   the provision of consistent temperature gradients,
(2)   the anisotropic nature of the materials, with different thermal properties in different directions,
(3)   the high thermal conductivity of metallic structures.

Some of these problems can be overcome with video-compatible infrared cameras and digital image processing, but thermographic methods on bond testing should be compared with alternative methods, such as mechanical impedance analysis with, for example, piezoelectric excitation and signal detection at low frequencies (1–10 kHz), and also with ultrasonic transmission techniques.

There are widespread applications of conventional thermography to items such as plate glass, automobile windscreens, non–metallic coating uniformity, plastic extrusions, electrolytic cells, and paper rolling and drying.

9.6 HOLOGRAPHY WITH NEUTRONS

If a specimen is placed in a neutron beam, the neutrons will be scattered throughout the volume, and not only on the surface: these neutrons can form a hologram, by placing an encoding device such as a Fresnel zone plate between the specimen and a neutron-sensitive film. The zone plate consists of a series of concentric circles which alternately absorb and transmit neutrons.

The radii of the zones, r_1, r_2, . . ., r_n, . . ., are such that $r_n = r_1 n^{1/2}$, where n is 1, 2, 3, . . ., N. The neutrons pass linearly (i.e. non-diffractively) through the zone plate and the method does not require spatial coherence of the neutron beam.

To reconstruct the image of any desired plane in the specimen, a coherent light beam produced by a transform lens is passed through a linearly-reduced transparency of the hologram and the undiffracted component is masked out. The reduced-size hologram must be on extremely-fine-grain film. Alternatively, if the hologram is digitised on, for example, a $512 \times 512 \times 5$ bit matrix, or is taken through a video-camera with a digital output, a mathematical transform technique will enable a computer to reconstruct the image; this image can also be digitally enhanced.

Suitable zone plates can be made by etching very thin gadolinium foil. The technique is still very new and image resolution is still comparatively poor, but, with a mini zone plate, can be as high as 1 mm. The image can be reconstructed to show any desired plane in the specimen, in terms of its neutron–absorption characteristics.

9.7   MICROWAVE METHODS

Electromagnetic waves at microwave frequencies can be used to measure dielectric properties or variations in these, and so are an important method of testing for such materials as insulators, ceramics, etc.

Microwaves have a wavelength within the range $10–10^{-3}$ m, so have a frequency in the range $3.10^9–3.10^{11}$ Hz. Conducting materials are opaque to radiation of this frequency so the method is applicable only to non–conducting materials or the surface of conductors. Short wavelength microwaves are similar in wavelength to the acoustic waves used in ultrasonic testing, and the size of detectable flaws ought therefore to be comparable.

A microwave generator is likely to be a klyston or magnetron with a waveguard coupling or antenna. The detector may be a silicon crystal, a detector 'horn' or antenna, cholesteric liquid crystals, or even a photographic film. The film is pre-exposed to even illumination, then coated with a layer of developer and placed in the microwaves. The heating effect of the microwaves affects the local development, and so the film density.

By varying the microwave frequency, using a sweep generator, the output signal minima can provide more information on the specimen, and this method has been used to detect flaws in ceramics.

Microwaves have been used for density testing of ultra-lightweight materials ($0.1–0.3$ g/cm^3), and in the form of sub-surface radar to detect buried cables.

Further reading: references

Bar-Cohen Y, A review of NDE of fibre-reinforced composite materials, *Mat Eval*, **44(4)**, 446, 1986

Cawley P and Adams R D, The sensitivity of the coin-tap method, *Proc 12th World Conf NDT (Amsterdam)*, **2**, 1223, Elsevier Press, 1989

Dean D S and Kerridge L A, Microwave techniques, in *Research Techniques in NDT*, Volume 1, Sharpe R S (ed), Academic Press, London, 1970

Florin C, Thermal testing methods, *Proc 4th Europ Conf NDT (London)*, **1**, 163, Pergamon Press, Oxford, 1988

ISO 3945:1977, *Mechanical vibration of large rotating machines with speeds in the range 10–200 revs/s. Measurement and evaluation of vibration severity, in situ*

Lange Yu V, Acoustic testing—low frequency acoustic methods, *Soviet J NDT*, **7/8**, 958, 1978

Magrab E B (ed), *Colloquium Papers on Vibration Testing*, ASME, New York, 1975

Non-destructive Testing Handbook (2E), Volume 1, *Leak Testing*, American Society of NDT, Columbus, Ohio, USA, 1982

Runkel J and Stegemann D, Non-destructive condition monitoring of nuclear power plants by vibration analysis, *Proc 12th World Conf NDT (Amsterdam)*, **2**, 628, Elsevier Press, 1989

Segal E and Rose J L, Chapter 8 of *Research Techniques in NDT*, Volume 4, Sharpe R S (ed), Academic Press, London, 1980

Sharpe R S (ed), *Research Techniques in NDT*, Volumes 1–8, Academic Press, 1980–86

10

Acceptance standards

10.1 INTRODUCTION

In all non-destructive testing, the findings obtained have eventually to be interpreted in terms of the suitability for service of the component or structure which has been inspected. Although this judgement should be based on much more information than non-destructive testing provides—e.g. materials, geometry, and surface-finish should be considered—and is the function of the designer, over the years it has been realised that the interpretation of non-destructive testing findings is a subject for the NDT expert, and many codes of acceptance/rejection have been written in terms of the flaws found by NDT. Most of these acceptance codes were written originally for radiographic inspection as this was the earliest NDT method which was capable of detecting flaws throughout the volume of a material. Also, as one of the most widespread early applications of radiography was to butt–weld inspection in steel, most of the codes concerned weld defects.

By radiography it is usually easy to identify the nature of the defect; its length and width can be measured on the film, but the third dimension (the through–thickness size) has to be estimated from the film density of the radiographic image, based on experience. There are special radiographic techniques, such as stereometry (see Chapter 3, Section 3.7.4), which will locate the position of the flaw in the specimen thickness, and also other special techniques (see Chapter 3, Section 3.7.5) which can measure the through–thickness dimension directly, but the latter are not well known or much used.

It is now well established from fracture mechanics considerations that the most serious defects from the loss of strength point of view are planar defects, such as cracks and lack of fusion, and that volumetric defects, such as gas holes, are less serious. It is clear that the through–thickness dimension of a defect is more significant than the defect length, and also that surface-breaking defects are more serious than totally internal defects. There is ample evidence that distributed porosity has very little effect on weld strength.

As a consequence of these findings, the acceptable levels for weld defects is a constantly changing topic, still subject to much discussion and controversy.

Many specifications, such as API 1104 (1980), still state acceptance require-ments in terms of defect length: for example, 'incomplete fusion at the root of the joint between weld metal and base metal shall not exceed 1″ in length, or

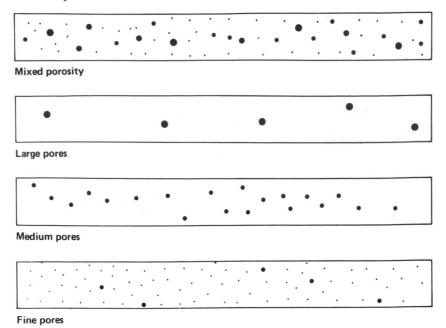

Mixed porosity

Large pores

Medium pores

Fine pores

Fig. 10.1 Porosity acceptance standards for butt-welds in pipelines for wall thicknesses over 12 mm. No individual pores should exceed 3 mm in diameter

more than 1″ in a total weld length of 12″; any elongated slag inclusions shall not exceed 2″ in length or $\frac{1}{16}$″ in width'. In this standard, the acceptability of distributed gas pockets is illustrated with diagrams for wall thickness less than $\frac{1}{2}$″, and for thicknesses over $\frac{1}{2}$″. An example is shown in Fig. 10.1. Similar porosity diagrams with different distributions are included in many other standards.

Over the years, codes of acceptance have ranged from 'Zero Defects Philosophy—no defects shown by NDT shall be acceptable', to more detailed codes, such as the API 1104 specification mentioned above. In most of the latter type of code, it is usually stated that no cracks or lack of fusion are acceptable, and specific dimensions are set for defects such as undercutting, lack of root penetration, distributed porosity, clustered porosity, linear porosity, slag inclusions, and a range of less common defects.

10.2   RADIOGRAPHY

Sometimes these acceptance codes are written in words, or numbers, with appropriate diagrams, but it is also convenient to illustrate the levels of spatially distributed defects, such as porosity, by a series of radiographs showing different degrees of severity. Such sets of radiographs are produced by the ASTM in the USA, and by the International Institute of Welding (IIW). Notionally, IIW *Collections of Reference Radiographs* are only for educational purposes, to illustrate the appearance of weld defects with different degrees of severity (up to five levels for many defects), and no acceptance levels are recommended by the IIW, but there is no reason why they should not be used to illustrate an acceptable level for any specific defect, for a particular application. The limitations of reference radiographs are as follows. Firstly, it has to be assumed that the radiographic image quality of the radiographs taken on a specific application is the same as the image quality of the reference radiographs. Secondly, the appearance of a defect in a weld depends markedly on the weld thickness, and it is not economically possible to produce reference radiographs for all metal thicknesses. Lastly, the appearance of planar defects also depends on radiation–beam orientation, so reference radiographs, while convenient, do not entirely remove the skill of the radiographic interpreter, and must be used with some discretion.

Obviously, the acceptability of a flaw must depend on the specific application of the welded structure and the final criterion must be agreed between the inspection authorities and the customer, but, as an example, the IIW has recently proposed a division of welded fabrications into the following three categories:

> moderate requirements (quality level D)
> medium requirements (quality level C)
> stringent requirements (quality level B).

The radiographic techniques to be used are divided into two classes, A and B, class B radiography requiring techniques having the better flaw sensitivity obtainable with fine-grain film, higher film density and a larger source-to-film distance. Class B radiography is required only for the stringent category of applications. The present proposals apply only to steel butt-welds in the thickness range 3–50 mm, and, *as an example*, contain the following proposals for the stringent category:

Flaw	*Proposed requirement*
Cracks	No visible cracks
Distributed porosity	Maximum 1% projected area, with none of diameter >3 mm or $0.2t$, where t = metal thickness
Clustered posority	Maximum 4%; diameter <2 mm or $0.2t$
Elongated cavities ⎫ Slag inclusions ⎭	Local defects permitted if width is less than 2 mm or $0.2t$
Lack of fusion ⎫ Lack of penetration ⎭	Not permitted
Undercut	Height not to exceed 0.4 mm
Excessive convexity	Height of reinforcement shall not exceed (1 mm + 10% of the weld width): maximum, 5 mm
Excessive penetration	Not to exceed (3 mm + 30% of the width of the penetration)

These are tentative and controversial proposals, meant to act as guidance to the contracting parties, and eventually to lead to unification of defect acceptance codes whether radiographic or ultrasonic weld testing is employed. They are for quality of workmanship, and are not 'fitness-for-purpose' criteria. Recently similar acceptance levels for weldments have been proposed in ISO/DIS 5817:1990, not limited to any particular non-destructive inspection method.

10.3   ULTRASONIC TESTING

In ultrasonic testing of welds, the situation is much more complex in that the nature of the defect is not always identified. Until recently, radiographic-type codes of acceptance have been used, but there have been attempts to redraft these in purely ultrasonic terms. Again, a distinction between quality-control and fitness-for-purpose criteria is made, and three classes, I to III, of inspection are proposed, class III being the most stringent. Fitness-for-purpose criteria are based on a fracture mechanics approach, which presupposes a knowledge of the applied stresses in service. For this approach, the through-thickness dimension of the defect is the critical parameter to be determined.

In ultrasonic testing, the following data are usually available:

(1)   the height of the peak signal,
(2)   the defect length, as measured by the length of probe movement parallel to the weld producing an indication greater than a specified amplitude,
(3)   the extent of probe movement perpendicular to the weld which produces an indication: this length of movement can be translated into a through-thickness dimension,
(4)   the number of defect indications per unit length, or unit volume of weld.

If the operator can distinguish between defect and non-defect indications, and can identify the nature of some defects either from their location, or from local probe movement (e.g. rotation), the above four criteria can be used to produce acceptance data. Figure 10.2 indicates the philosophy proposed.

The ultrasonic equipment is calibrated: the evaluation and rejection thresholds AA' and BB' are set according to agreed criteria. When a defect is found, and proved to be real, the probe is adjusted for maximum signal. If the amplitude is below BB', the defect is accepted; if it is above AA', the defect is recorded as rejectable. If the amplitude is between AA' and BB', the defect is examined in more detail to decide whether it is cracklike; it is checked for length, or number/unit length and compared with agreed criteria. Line CC' in Fig. 10.2 is the amplitude reference level established by reference to standard reflectors, such as side-drilled holes or flat-bottomed holes.

In fitness-for-purpose examinations, a similar procedure has been proposed, but with more emphasis on measuring defect size, by starting with a higher sensitivity (e.g. 'grass' just visible), more detailed scanning over any defect found, and dimensioning by means of a distance–gain–size diagram, or echodynamic evaluation.

A proposed acceptance criteria for the three classes of work is shown in Table 10.1, for manual ultrasonic inspection, for longitudinal weld defects. If there are more than four measurable defects in a length equal to four times the wall thickness, the sensitivity should be increased by 4 dB and the inspection repeated. These proposals for an ultrasonic inspection standard also cover the extent of the inspection, the number of probe angles to be used, the surface condition, and a guide to defect characterisation into four types of echodynamic pattern.

An alternative, simpler approach to specifying acceptance standards for ultrasonic weld inspection is simply to state that no defects which occupy more than a specific percentage of the weld thickness shall be acceptable: for example, 20% of the nominal thickness for stringent applications, or 30% for

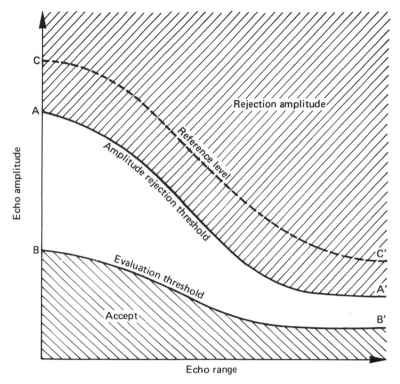

Fig. 10.2 Principles of a proposed acceptance standard for ultrasonic weld testing, based on quality-control criteria. The curves are determined for specific applications

moderately-stressed applications. Such an approach has been proposed in ISO draft DP 5817, but such simplification is in effect ignoring a lot of data which the ultrasonic signal contains, and should be supplemented by the criteria for individual types of defect as detailed above for a radiographic acceptance standard.

Table 10.1 IIW proposals for defect acceptance by ultrasonic examination

Quantity	Class, thickness range (mm)		
	I, 7–50	II, 7–100	III, 7–300
Evaluation theshold	DAC + 10 dB (33% DAC)	DAC + 14 dB (20% DAC)	DAC + 14 dB (20% DAC)
Amplitude rejection level	DAC	DAC + 4 dB (65% DAC)	DAC + 4 dB (65% DAC)
Length evaluation level	DAC + 10 dB	DAC + 10 dB	DAC + 10 dB
Maximum acceptable length (if above length evaluation level)	Wall thickness	75% wall thickness	50% wall thickness

Notes
DAC – distance–amplitude correction value for the distance of the echo from the probe. The DAC curve is established with 3 mm diameter drilled holes. (DAC + 10 dB) means a level 10 dB more sensitive than a 3 mm cylindrical hole at the range of the defect.

It is worth noting that data handling techniques are developing so rapidly in ultrasonic testing that these acceptance criteria might eventually have to be reformulated in a digital data format, to enable the acceptable sizes of defects to be built into a pattern–recognition program.

10.4 CASTINGS

Reference radiographs for steel castings were first issued by the US Bureau of Ships in 1942 and this collection was later adopted by the ASTM as the E71 Reference Radiographs. They covered steel castings up to 2″ thick, and illustrated the main defects found in castings: gas and blowholes, sand spots and inclusions, internal shrinkage, hot-tears, cracks, unfused chaplets, and internal chills. There are four, five, or six degrees of severity of each defect, illustrated with an X-radiograph and a gamma-radiograph, each 17 × 12 cm. Castings are divided into five categories and each radiograph is indicated as acceptable, borderline, or unacceptable for each category of casting.

Since this first reference radiograph collection, the ASTM has published further sets for aluminium and magnesium castings, aluminium and magnesium die-castings, high-strength copper base and Ni–Cu castings, tin–bronze castings, thin steel precision castings, heavy-wall steel castings, up to 12″, and grey iron and ductile iron castings. Some of these also cover a range of radiographic sources.

The use of reference radiographs is a convenient way of grading the defects found in castings, but has the same weakness as the use of reference radiographs of welds, in that a competent designer must state acceptance levels *for each particular application* in terms of the standard radiographs available. In castings, there is also a second complication in that the common place for serious defects is at a section-change, and this is not illustrated by the present collections.

10.5 NEUTRON RADIOGRAPHY

A collection of reference neutron radiographs of nuclear reactor fuel has been produced by a Neutron Radiography Working Party (EUR/8916EN/EP/ 1984). These show the defects and abnormalities which can occur in nuclear fuel pin components for light-water and fast reactor fuel elements.

Further reading

API Specification 5LX, *Specification for High Test Line Pipe* (23E), American Petroleum Institute, Washington DC, 1980
API Standard 1104 (15E), *Standard for Welding Pipelines and Related Facilities*, American Petroleum Institute, Washington DC, 1980
BS:4515:1984, *Specification for the Process of Welding of Steel Pipelines on Land and Offshore*, British Standards Institution, London
Collections of Reference Radiographs for Butt-welds in Steel, The International Institute of Welding, London

PD:6493:1980, *Guidance on Some Methods for the Derivation of Acceptance Levels for Defects in Fusion Welded Joints*, British Standards Institution, London

Reference Neutron Radiographs of Nuclear Reactor Fuel, D Reidel Publishing Co, Dordrecht, Boston and Lancaster, 1984

Reference Radiographs for Welds and Castings, ASTM, Philadelphia

11

Reliability and probability

In recent years it has been realised that even when it can be shown that a specific NDT technique has the ability to detect a given size of flaw, this does not necessarily mean that the flaw will always be detected, and several studies of probability and realiability have been initiated.

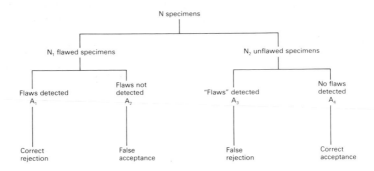

Fig. 11.1 The four possibilities of inspection. Total number of specimens, $N = N_1 + N_2$

Assuming N specimens are tested, Fig. 11.1 shows the possibilities. There are four possible outcomes of the inspection, and the 'probability of detection (POD)' is given by A_1/N_1, with the 'probability of recognition (POR)' $= A_4/N_2$. The false call probability is A_3/N_2. 'Relative Operating Characteristic (ROC)' curves plot (POD) against 'false call probability (FCP)'.

A typical ROC curve is shown in Fig. 11.2. Curve A represents high quality inspection, curve B is lower quality, and line C is the 'chance' line. If the ROC curve is plotted on double probability paper, straight-line plots are obtained. To develop a realistic ROC curve a large number of specimens and readings are necessary on materials where the existence or otherwise of the particular flaws has been confirmed.

The accuracy of the observer is defined as $(A_1 + A_4)/N$, and the sensitivity of detection as $A_1/N_1 = POD$.

ROC curves have not yet been widely used in non–destructive testing due to

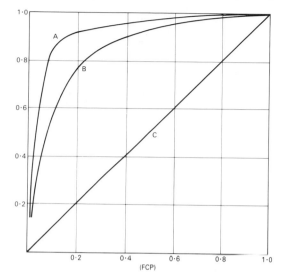

Fig. 11.2   ROC curves showing very good (A), good (B) and 'chance' (C) curves

the difficulty of amassing sufficient experimental data, but in theory at least they can be used to compare techniques, assess different image enhancement systems, or compare operator performances.

Plots of POD against defect detection size for various NDT methods have been derived experimentally (Kauppinen, 1989), and are perhaps more easy to interpret. An example, due to Kauppinen, is shown in Fig. 11.3.

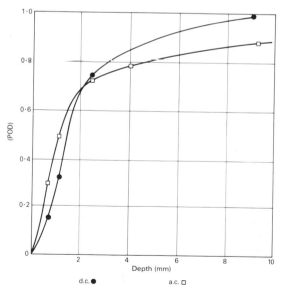

Fig. 11.3   The effect of magnetising current on flaw detectability in MPI (after Kauppinen and Sillanpää)

It is also possible to calculate the number of tests which are needed to achieve a particular confidence level for a chosen POD. Thus, according to Stone (1987), for a POD of 90% and a confidence level of 95%, in 29 trials on 58 samples all 29 defective samples must be detected: if one defect is missed then 46 trials must be made with no further misses to achieve the same confidence level. This would require 92 samples to be inspected. These are the number of samples for each type of defect and each range of defect size: the need is therefore for a lot of samples. Nevertheless POD curves have been derived for some NDT techniques (Fig. 11.4), based on several thousand observations. Figure 11.4 shows POD curves for an ultrasonic test at a particular sensitivity level (20% DAC from a 3 mm side-drilled hole) based on all types of weld defect in 25 mm steel. The curve is the 95% confidence band from a fitted regression curve.

The practical limitations of ROC curves in NDT seems to be the need for very large quantities of experimental data and the need for data where only one parameter is varied at a time. In most NDT methods there are a large number of parameters affecting flaw sensitivity, and even for any specific flaw, there are several dimensional parameters. Thus, in radiography, if specimens are chosen to assess crack sensitivity and the cracks are all 'tight', a result will be obtained which shows that radiography has a very poor performance: on a wider range of cracks a totally different conclusion would be reached.

Silk and others (1987) have investigated the performance of various NDT methods from a different point-of-view. The flaw size distribution in a particular type of structure has been estimated and the desirable sizing precision of various NDT techniques assessed.

A perfect technique would isolate and size all the flaws above a defined size and produce no false calls. However, if the discrimination level is set precisely

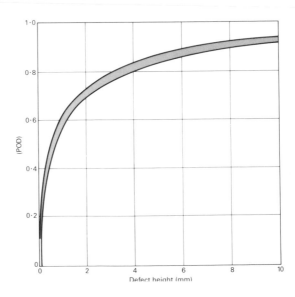

Fig. 11.4 POD against defect height, for an ultrasonic test on welded steel butt-welds (after Forli and Pettersen). The curve is the 95% confidence band of a fitted regression curve

at the 'acceptable flaw' level some larger flaws will inevitably be wrongly sized and be accepted, and vice-versa, so normally the discrimination level is set lower, so that fewer significant flaws are missed, and this results in an increase in false calls. In addition to equipment performance, an assessment of human reliability suggests a failure rate of about one in 2000.

In any NDT technique there are random errors and systematic errors. The systematic errors are usually dominant and are basically the limitations of the technique, as distinct from the random errors when the same flaw is measured several times. As examples of the results obtained, recent experimental work quoted by Silk (1990), has shown that with conventional ultrasonic flaw sizing methods (dB drop) the errors range from 3–5 mm, whereas a calibrated ultrasonic time-of-flight method can reduce this to 0.5–0.7 mm, and an a.c. potential drop method to 1–2 mm.

The basic causes of these systematic errors are still to be studied.

References

Forli O and Pettersen B, The unreliability of non-destructive examinations, *Proc 4th Europ Conf NDT (London)*, **2**, 833, Pergamon Press, Oxford, 1988

Kauppinen P and Sillanpää J, *Proc 12th World Conf NDT (Amsterdam)*, **2**, 1723, Elsevier Press, 1989

Silk M G, Stoneham A M and Temple J A G, *The Reliability of NDT*, Adam Hilger, Bristol, 1987

Silk M G, Whapham A D and Hobbs C P, *Proc NDT-89 (Sheffield)*, **1**, 3, Pergamon Press, Oxford, 1990

Stone D E W, The reliability of inspection techniques for military aircraft, *Proc 4th Europ Conf NDT (London)*, **2**, 819, Pergamon Press, 1988

Swets, J A, Assessment of NDT systems, *Mat Eval*, **41(11)**, 1294, 1983

Taylor T T, Haesler P G and Doctor S R, Use of ROC curves in measuring inspection performances, *Proc 12th World Conf NDT (Amsterdam)*, **2**, 1059, Elsevier Press, 1989

Todd-Pokropek A, Physical Aspects of Medical imaging, in *ROC Analysis*, John Wiley and Sons, London, 1981

12

Computers in NDT

The extremely rapid evolution of computer technology has brought computers into almost every field of NDT. Small, fast and powerful data processors are now commercially available at reasonable cost. The great increase in power and availability of computers has caused an enormous increase in the amount of data which can be handled, and although computer graphics has complemented this expansion there are still problems in assimilating the mass of data which can be produced, particularly in ultrasonic and eddy current testing. Advanced programming methods such as adaptive learning, expert systems and neural networks are now being talked of as practical procedures, with potential applications in the NDT field.

There are many benefits from the use of computers, from the development of automated systems to improve the realiability and capability of NDT, to increased data storage capacity and extended data analysis techniques leading to more rapid inspection methods. Where a probe has to be positioned on the test specimen in such NDT methods as ultrasonic or eddy current testing, a computer program can be used to manipulate the probe, with an indexing function to measure and record the probe position and orientation. Alternatively, a smaller specimen can be positioned under a probe by a robot handler working under computer control. Various indexing devices are possible— potentiometer, incremental encoder, synchro-resolvers used in pairs. A second computer program can be used to set up the parameters of the test procedure—sensitivity, probe range, probe angle, etc., but the handling of the acquired test data, probably the most important procedure, is the field in which progress has been the most rapid. Taking ultrasonic flaw detection as an example, the received pulse from a flaw can be measured digitally in both range and amplitude. In addition, the pulse shape can be recorded with various degrees of detail: this is so-called 'raw data' and can be collected from one or more transducers. This data can then be analysed to produce the apparent location and dimensions of the flaw, and presented either digitally or graphically. An important function of the computer program is to compress the data, discard the unnecessary items ('data reduction') and present an improved 'picture' of the flaws. The flaw found can then be evaluated in terms of a built-in program covering flaw length, through-thickness dimension, aspect ratio, etc.

With a non-destructive technique which images the flaws, such as radiography or magnetic particle inspection, the image can be digitised by dividing it into picture-elements (pixels) and again storing and processing this data in digital form. The source of the data may be a pre-produced image (film, screen, magnetic particle indications or penetrant indications on a surface), a television camera image, or a scan with an array of detectors. Again, the purpose of the computer may be simply to store and re-present the image to improve the image by digital enhancement programs, to compare with a standard image, or to evaluate in terms of some programmed standard of flaw-length, etc.

As described so far, these might be classed as 'rule-based systems', but it is feasible to proceed much further into so-called 'artificial intelligence' and 'expert' systems. The shell of an expert system is:

(1)   The knowledge base—what is acceptable/rejectable in terms of flaw size, nature, location.
(2)   A means of applying this knowledge to make useful inferences on the data collected—the inference engine: a data interface is obviously essential.
(3)   A user interface to communicate the inferences to the user.

In an inductive system a number of examples are inputted and the expert system then deduces the rules. There are obvious problems when the examples are not sufficiently comprehensive, or the programmer and expert are not the same person. An expert program language has been found to be essential, e.g. PROLOG or LISP. Conventional program languages such as FORTRAN and BASIC have been found to be inadequate for this purpose. TIMM and EXPERTEASE are knowledge acquisition tools which can form their own rules by rule induction. SAVOIR is an expert system shell written in PASCAL, which provides an interface to FORTRAN.

In conventional procedural programming one starts at a point A, and by a series of steps one arrives at an answer B. In an expert system one starts by asking what is the answer B and then by examining the available facts and rules in the knowledge base, one determines what data is required to produce this answer. The necessary questions to acquire this data will be asked and another derivation for the answer will be attempted if the first route fails. The means by which the system carries out these operations is called the inference engine. An expert system can never be better than the experts who provide the knowledge base, but it has advantages in speed and consistency in applying the rules. It can also work on an amalgam of knowledge derived from several experts.

Such expert systems are beginning to be applied in the fields of automated ultrasonic flaw detection and acoustic emission, where a multiplicity of transducers produces very large quantities of data. At the present time they are used mostly as filters to eliminate non-significant indications, leaving a small number of suspect indications for evaluation by a human expert, but this is probably only an interim stage before fully automatic evaluation. Rigorous system validation is necessary for consistency, completeness, soundness, precision and usability. An expert system necessarily operates in a very narrow problem domain. It is still necessary to measure the performance of an expert system by comparing it with the performance of human experts.

There is now considerable literature on what are known as computing neural networks. Neural networks mimic the method by which the human brain is understood to operate, by using a network of processes which represent nerve-cells (neurons). Using three layers, with each neuron linked to those in adjacent layers, the outer layers act as inputs and outputs. The links are adjusted in weight by software as the network 'learns' a particular task. Instead of performing a program of instructions sequentially, a neural network explores many hypotheses simultaneously, using massive parallel networks. Neural networks may therefore be a powerful computer technique for pattern recognition and pattern classification.

In a digital data system which constructs a two-dimensional image, such as an X-ray image, computers are used mainly for the purpose of enhancing the image (i.e. improving its quality). If image quality is described in terms of contrast, sharpness and noise, simple programs are available for contrast enhancement both overall and local, for field equalisation (field flattening), and for noise reduction (by averaging a number of television frames). Histogram equalisation is a powerful method of contrast adjustment. The pixel values over a limited range, are expanded to fill the full grey level range, or thresholded as necessary. Pseudo-colour can be added and in some cases is claimed to enhance contrast perception. Image sharpening is a more complex problem but enhanced images can be obtained by convolution calculations based on a 3×3, 5×5 or 9×9 pixel filter. As an example the values chosen in the filter

-1	-1	-1
-1	8	-1
-1	-1	-1

are applied by multiplication to each block of nine pixel values in the image to give a new value for the central pixel. These values are integrated over the whole image plane to produce an enhanced image.

Some computer operations are more conveniently made if the image is transformed into the spatial frequency domain, and this can be conveniently done using the Discrete Fourier Transform (DFT). If the higher spatial frequencies are attenuated by multiplying the pixel values by a suitable low-pass filter function and then the DFT is inverted, a smoother image will be obtained. Similarly if a high-pass filter is used, a sharper but more noisy image is possible. The removal of specific spatial frequency components from a two-dimensional Fourier Transform, followed by transformation of the image back into the image plane, can produce important enhancement effects which do not yet seem to have been fully explored on X-ray images.

The Laplacian operator can be combined with the original image to produce edge enhancement, which has the effect of making the image appear crisper. A similar effect can be produced by taking the first derivative of the image function.

A very large amount of literature on digital image enhancement is now in existence, and new and more complex procedures for image improvement are regularly proposed. Most of these papers consider optical images and the correction of optical degradations, but some of the proposed techniques should also be applicable in the radiographic field.

A further use of computer software is for pattern recognition procedures. Taking the case of a weld as an example, there are a limited number of possible weld flaws, and if each of these can be specified by a number of parameters—length, width, intensity distribution, contrast, etc. the computer can be used to compare the parameters of a detected flaw in the image with these, and so specify the likely nature and size of the flaw. In turn, this set of parameter values can be compared with those of acceptable and rejectable flaws, and an automatic accept/reject interpretation system can be programmed. Alternatively the computer data store can hold the parameters of accept/reject images from a collection of reference radiographs and compare a detected flaw with this catalogue and so produce a probability-of-identification table. As a simple example: If the length/width ratio of a detected centre-line flaw image is greater than (say) 10, the flaw must be either a crack or lack of penetration, and would for many applications be rejectable without more precise identification.

The present practical limit to such pattern recognition techniques is partly due to the masking effects of background variations in the image (which can be overcome), and to the image scanning systems used, which are inherently more responsive to image noise than the human eye. In layman's terms, at present the skilled eye can discern and interpret lower contrast images than the microdensitometer or array producing the basic image data for the computer.

In eddy current testing, computers have rapidly come into widespread use. A microprocessor can be used to:

(1)   control the instrument calibration, store readings, improve data, etc.;
(2)   produce graphical displays, recordings, hard copy—i.e. communicate with the operator;
(3)   develop intelligence operations and decision-making procedures.

Most eddy current testing at present uses a single probe, and there is a screen display which is not in any sense a pictorial image of the flaw, but scanning techniques or arrays of detectors can be used, and then it is feasible to programme a computer to produce a flaw image.

In acoustic emission methods a computer is virtually essential to handle the enormous amount of data generated, to condense and analyse this and to produce graphical displays and print-outs.

As a final example: An X-ray fluorescence analysis instrument is marketed which uses one of two radioisotopes (Cd^{109}, Fe^{55}) as the radiation source; a mercuric iodide radiation detector detects the characteristic X-rays from a specimen in an energy-proportional mode, and analyses these for any of 21 elements. In addition, the microprocessor holds the spectra for 225 specific alloys and can compare the received spectra with these. The whole system is small enough to be field-portable and a reading takes about 30 s to obtain and process.

Another field in which computers have enabled rapid developments to be achieved is in 'modelling'. This applies particularly to magnetic, eddy current and ultrasonic techniques. Modelling techniques can range from a mathematical simulation of elastic wave propagation in an elastic solid containing interfaces and stylised flaws to determine the wave-shapes resulting from the flaws, to more simple simulations of the beam characteristics of various types of focussing probe. The most common method for electrical and magnetic fields is the numerical 'finite element' method in which the material is divided into simple shaped cells and a function is evaluated for each cell which represents the electromagnetic energy in that cell. The total energy in the material, probe and surround is obtained by summing over the cells.

Mathematical ray-tracing has been used to determine the ultrasonic beam pattern through an austenitic steel weld.

Further reading

Becker R, Betzold K, Boness K D, Collins R, Holt C C and Simkin J, The modelling of electrical current NDT methods, *Brit J NDT*, **28**, 286 and 361, 1986

Coffey J M, Mathematical modelling in NDT, *Proc 4th Europ Conf NDT (London)*, **1**, 79, Pergamon Press, Oxford, 1988

McNab A, Young H and Durranti T S, An expert system for NDT ultrasonics, *Proc 4th Europ Conf NDT (London)*, **1**, 551, Pergamon Press, 1988

Index